PHYSICS EXPERIMEN
AND PROJECTS FOR
STUDENTS

Vol. III

PHYSICS EXPERIMENTS AND PROJECTS FOR STUDENTS

Vol. III

Edited by

C. Isenberg
University of Kent at Canterbury

and

S. Chomet
King's College London

USA	Publishing Office:	Taylor & Francis 1101 Vermont Avenue, N.W. Suite 200 Washington, DC 20005-3521 Tel: (202) 289-2174 Fax: (202) 289-3665
	Distribution Center:	Taylor & Francis 1900 Frost Road Suite 101 Bristol, PA 19007-1598 Tel: (215) 785-5800 Fax: (215) 785-5515
UK		Taylor & Francis Ltd. 1 Gunpowder Square London EC4A 3DE Tel: +44 (0) 171 583 0490 Fax: +44 (0) 171 583 0581

PHYSICS EXPERIMENTS AND PROJECTS FOR STUDENTS, Volume III

1 2 3 4 5 6 7 8 9 0 BCBC 9 8 7 6

A CIP catalog record for this book is available from the British Library.
∞ The paper in this publication meets the requirements of the ANSI Standard Z39.48-1984 (Permanence of Paper)

ISBN 1-56032-280-2

Contents

Computing and Nuclear Physics

Preface

Throughout the country there is a considerable number of undergraduate physics laboratories. The experiments in these laboratories have been developed by the local staff, often making use of the specialized interests of the department. There has been little interaction between academics who are responsible for these laboratories. Consequently I suggested to the Education Group of The Institute of Physics that it should organize and exhibition containing some of the standard teaching experiments available in undergraduate laboratories at universities, polytechnics, teachers' training colleges, and institutes of higher education. This was subsequently extended to include advanced school projects. Such an exhibition would enable institutions to exchange information and ideas concerning experiments and enable new experiments to be introduced into laboratories without the considerable effort that is required to develop a new laboratory experiment.

The Education Group welcomed this suggestion. The first such exhibition was held in March 1984 and was a considerable success, attracting academic staff from all over Britain together with teachers and pupils from London schools. This success has encouraged us to organize future exhibitions of this nature and to publish this series of monographs.

This is the third volume in the series and it presents experiments and projects from the third exhibition that was held at The Institute of Physics in September 1987.

Cyril Isenberg
University of Kent at Canterbury

Nonpropagating solitons

J.G.M. ARMITAGE AND J.F. ALLAN
Department of Physics, University of St. Andrews, North Haugh, St. Andrews, Fife, KY16 9SS

A 'solitary wave' was first reported in 1843 by John Scott Russell[1] who observed one on the Union Canal near Edinburgh as a result of the movement of a boat in a narrow section of the canal. He noticed that a localised 'wave' was created and traveled along the canal to a great distance without any change of form. Normally, a disturbance on the surface of water is made up of a number of waves of different wavelength, and since these have different wave velocities, the resultant disturbance fairly rapidly changes shape and disperses. Scott Russell's wave was one which could not be described by the simple linear wave equation. Such a localized wave is now referred to as a 'soliton' and belongs to a class of solutions to an extensive range of nonlinear partial differential equations.

A type of soliton was found by Wu, Keolian and Rudnick.[2,3] It took the form of a localized back and forth mode across the width of a long narrow tank, with the surface to either side of the wave practically motionless (Fig. 1). To maintain the soliton the tank has to be continuously vibrated. Such a soliton is very stable and allows us to investigate some of the general properties of solitons.

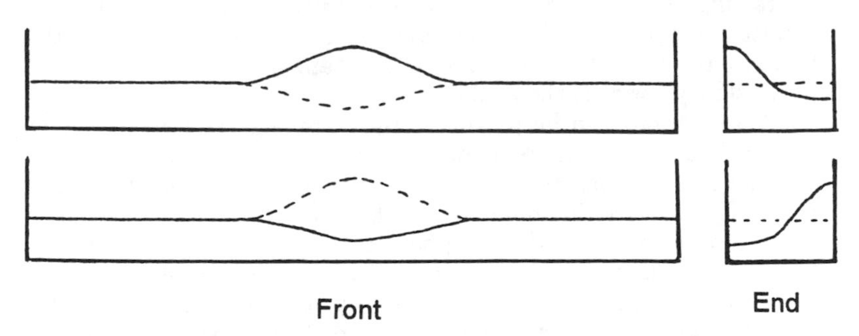

FIG.1 Views of the soliton separated by half a period

Various resonant modes of surface waves can be observed when the frequency of vibration produces waves whose wavelength satisfies the condition

$$\lambda = \frac{2L}{n} \text{ or } \frac{2W}{n} \tag{1}$$

where L is the length of the tank, W the width, and n an integer. The velocity c of these surface waves[4] in water of depth d is given by

$$c^2 = (\lambda g/2\pi)\tanh(2\pi d/\lambda) \tag{2}$$

where g is the acceleration of gravity.

The lowest mode for a wave transverse to the tank is unstable against the formation of solitons. A single soliton can be produced by reducing the frequency by a few percent from that calculated from (1) and (2). The wavelength should thus be slightly too long to fit across the tank ($\lambda/2 > W$). However, because the equations are nonlinear, the velocity of the wave depends slightly on amplitude and a wave at this frequency, but with a large enough amplitude, can satisfy the condition $\lambda/2 = W$.

A wave is localized to part of the length of the tank where the amplitude is large enough for the boundary conditions to be satisfied, but prevented from spreading sideways into regions where the amplitude is small and where the boundary conditions cannot be satisfied. The soliton exists because of the appropriate balance between the dispersive and nonlinear characters of the surface wave.

Solitons are only stable for a limited range of drive amplitudes, frequencies, and depths of water. They are repelled by a slight narrowing of the width, so they can be pushed along by a piece of plastic a couple of millimeters thick which is slid along the wall. Shallow regions attract them so if the tank is not level they will move slowly towards the shallow end.

These solitons can be produced by vibrating the tank horizontally and at the same frequency as the wave, but this limits observation to solitons that are all in phase. By driving the tank vertically, at *twice* the frequency of the wave, it is possible to have two or more solitons either in phase or 180° out of phase. In the former case they will attract and interact in a complicated way before (i) collapsing to one, (ii) coming together separating into two and then repeating this motion indefinitely or (iii) breaking up into more solitons. Solitons that are out of phase repel.

Equipment

The tank is made from glued Perspex sheet,72 cm × 2.9 cm × 7.5 cm, and contains water to a depth of 1.5−−2.5 cm (Fig. 2). The length is not critical and 34 cm works very well. The tank is suspended so as to allow pure vertical motion, but is rigid in all other directions. This is achieved by using two flat parallel copper-nickel springs at one end and a similar one at the other end. The use of only three springs ensures that the motion is not over-constrained. The weight of the tank is supported by a compression spring concentric with the vibrator whose actuator is lightly spring loaded on to the base of the tank. Since the flat springs support none of the weight they can be made from any reasonably stiff metal or plastic foil about 0.2 mm thick. This method of suspension is virtually frictionless and so provides a minimum load for the vibrator. The resonant frequency of the tank on its suspension is about $7\,Hz$, slightly below the $10\,Hz$ needed to drive the soliton. This worked well and no attempt has been made to optimize it. The vibrator is an Advance VI, and requires a current of 0.15 A into 3.6 Ω to maintain a soliton, and up to 0.4 A to initiate one.

The drive amplitude ($\sim$ 0.25 mm) is monitored very simply by looking at the off-balance signal of a small capacitor in a bridge circuit. The fixed parallel plate capacitor is varied by the movement of a grounded vane between the plates. The amplitude of the wave ($\sim$ 10 mm) is measured by the current which flows between two thin vertical wire electrodes dipping into the liquid. The current is driven by an a.c. signal to avoid electrolysis. (Fig. 3). The current is a reasonably linear function of the wetted depth of the electrodes.

Generation of solitons is facilitated by reducing the depth of water to about a half over a few centimeters at each end of the tank. The effect is to increase dissipation to the longitudinal wave modes while not affecting the transverse modes and therefore preferentially selecting the required modes. In fact if sloping beaches are used fairly close together it is possible to stabilise the soliton over a wider range of frequency and drive amplitude and even to obtain a higher mode with the wavelength equal to the width of the tank. For this the shorter tank was used and the depth of water reduced. A mode with one-and-a-half wavelengths has not been clearly seen, and it is thought that surface tension is playing a significant role apart from the possibility that it may not exist.

There can be problems with the water not wetting the walls. This affects the shape and stability of the soliton, which is cured

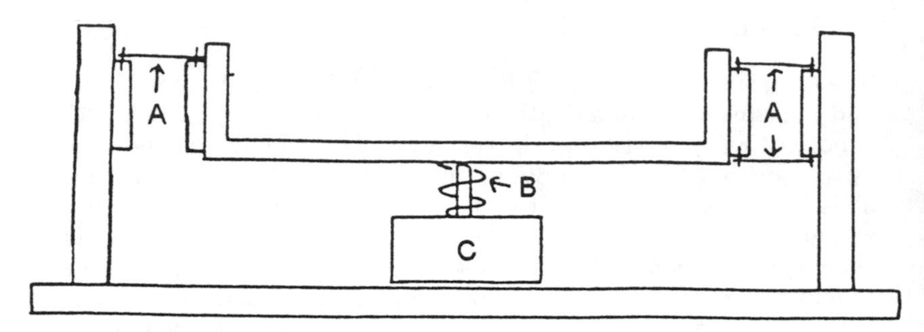

FIG.2 Tank mounting: A – parallel flat springs, B – support spring, C – vibrator

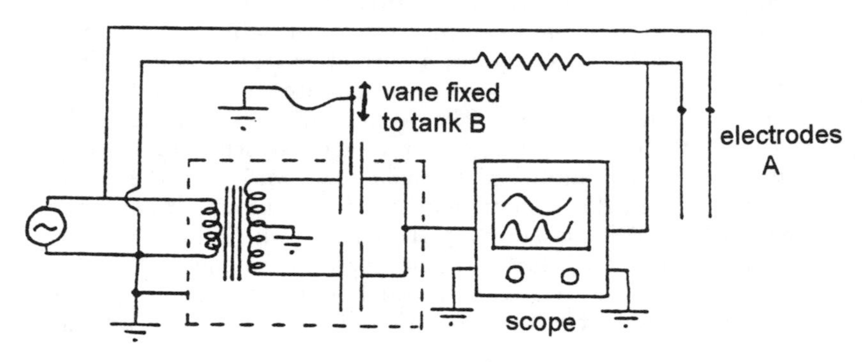

FIG.3 Wave and tank amplitude detectors A and B, respectively

by the addition of a few drops of a good photographic wetting agent.

Theory

A third order perturbation expansion has been carried out by Larraza and Putterman[5] (see also Ref. 6) who found that the appropriate equation was the nonlinear Schrödinger equation. In the lowest order, the profile of the wave is

$$\psi = A\,\mathrm{sech}(\alpha x)\cos(\pi n y/W)\sin(\omega t)$$

where A and α are complicated functions of ω, and x and y are the coordinates along and across the tank, respectively.

The theory gives a good qualitative description of the behavior, but the amplitudes of the solitons produced are large and better approximations would be required for quantitative agreement.

Experiments

The following are some suggestions which could be fairly readily investigated:

(1) the ranges of frequency and amplitude for which one or more solitons are stable
(2) dependence of the profile on frequency
(3) the effects of surface tension and viscosity[6]
(4) periodicity of two attracting solitons.
(5) pinning forces by locally altering the width and depth of the tank.
(6) dependence of wave velocity on amplitude.[3]

References

1 J. Scott Russell, 'Report on Waves', *Br. Assoc. Adv. Sci. Rep.* **14**, 331 (1844)
2 J. Wu, R. Keolian, and I. Rudnick, 'Observation of a Nonpropagating Hydrodynamic Soliton', *Phys. Rev. Lett.* **52**, 1421-4, (1984)
3 J. Wu and I. Rudnick, 'Amplitude-Dependent Properties of a Hydrodynamic Soliton' *Phys. Rev. Lett.* **55**, 204-7 (1985)
4 L. D. Landau and E. M. Lifshitz, *Fluid Mechanics*, Pergamon, London (1959)
5 A. Larraza and S. Putterman, 'Theory of Non-Propagating Surface-Wave Solitons', *J. Fluid Mech.* **148**, 443-9 (1984)
6 J. W. Miles, 'Parametrically Excited Solitary Waves', *J. Fluid Mech.* **148**, 451-60 (1984)

Chaos in the laboratory

S.J. ROGERS, M.R. HALSE, J.S. FOWLER, AND T.S. ROUSE
The Physics Laboratory, University of Kent at Canterbury, Canterbury, Kent, CT2 7NR, UK

Abstract: This third year undergraduate experiment is concerned with the behavior of a simple nonlinear circuit consisting of an inductor, a diode, and a resistor in series. The response of the circuit is studied for drive voltages in the range 0 – 8 V peak-to-peak at frequencies in the range 100 kHz to 1 MHz. Chaos is seen in the relationship between the current and voltage, which is governed by both the amplitude and the frequency of the input. Depending on these control parameters, the current waveform may be a simple sine wave at the drive frequency or may contain more subharmonic components; chaos arises when the number of subharmonic frequencies tends to infinity. The chaotic behavior is successfully modeled using FAST BASIC on an Atari ST computer.

Introduction

Chaos, as the state out of which the world was created, is a religious idea of very ancient origin, predating even the book of Genesis, but the importance of chaos as a scientific and mathematical concept has been widely recognized only recently.[1] Since the time of Newton, physicists have used time-dependent equations to predict the behavior of systems, in which cause and effect are well understood, in the belief that the present determines the future. It has become increasingly clear, however, that this is by no means always the case in systems in which nonlinear effects play a role. It may be impossible to predict the time evolution of such systems, which are then characterized as *chaotic.* A system may exhibit chaos or act predictably, depending on the values of the parameters by which it is characterized. In the situation of chaos, vanishingly small differences in the initial condition of the system lead to totally different patterns of future behavior.[2,3]

The early realisation by biologists that, in some circumstances, the quasirandom fluctuations in wildlife populations could be modelled by simple nonlinear equations, such as the logistic equation discussed below, was paralleled by developments in mathematical topology and by a growing awareness that similar effects are to be seen in a wide range of physical systems from the weather to the dripping of a tap.[4] The work presented here is concerned with such a system, an LCR circuit in which the capacitive element is a varactor diode. With some embellishment, it represents the third year undergraduate laboratory project work of two of us (JF and TR). The LCR circuit is driven by an oscillator at frequencies typically in the range $0.1 - 1\,\mathrm{MHz}$. Chaos is seen in the relationship between the phase and amplitude of the driving voltage and of the corresponding current in the circuit. The first step towards chaos is seen in the appearance in the current of the sub-harmonic of the drive frequency.

The logistic equation

Many of the features that characterise chaotic systems are seen in the solutions of the so-called logistic equation. It generates iteratively a series of real numbers in which the $(n+1)$th value, $x_n + 1$, is given in terms of the nth value, x_n, by the nonlinear relation

$$x_n + 1 = \lambda x_n(1 - x_n), \ldots \tag{1}$$

where λ, the control parameter that determines the characteristic form of the series, must lie in the range 0 to 4 if the series is to remain finite.

For $\lambda < 3$ the iteration tends to a single limiting value which is the solution of the equation

$$x = \lambda x(1 - x)$$

The way in which the series approaches this limit, graphically the point of intersection of the two equations

$$y = \lambda x(1 - x)$$

and

$$y = x$$

is easily seen in Figs. 1a and 1b. Starting with an arbitrary choice x_1, x_2 is the corresponding y value on the parabola, which, in turn, yields the next iteration by 'reflection' from the 45° line, $y = x$.

If $\lambda < 1$, the parabola is always below the 45° line and the iteration works towards the trivial solution $x = 0$. For $1 < \lambda < 3$,

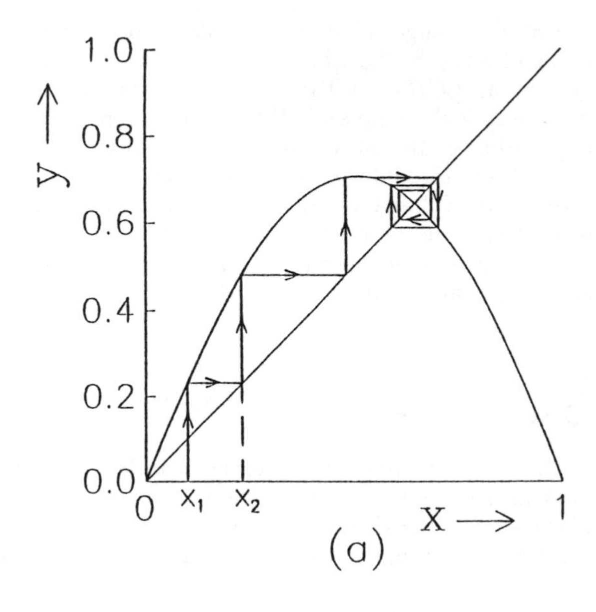

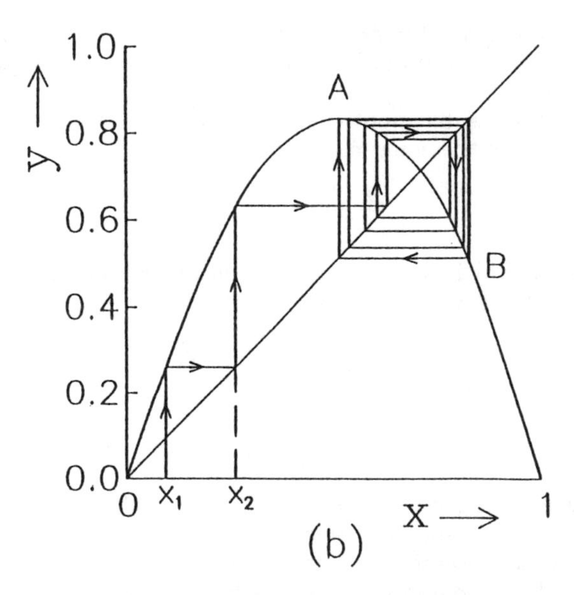

FIG.1 Graphical iteration of the logistic equation (1) for two values of the control parameter: (a) $\lambda = 2.8$, (b) $\lambda = 3.3$

the modulus of the slope of the parabola at the point of intersection of the two curves is less than 1 and the graphical construction converges to this point, as indicated in Fig. 1a. As λ increases beyond the value 3, at the common point the parabola becomes steeper than 45° and the construction stabilises to a rectangular pattern, as shown in Fig. 1b, which corresponds to the series alternating between the values at A and B; the transition to two values is termed *bifurcation*. Further increase in λ leads first to a situation in which the iterative series takes four values cyclically, corresponding to two points on either side of the maximum in the parabola, and, after an infinite series of further bifurcations, this in turn gives place to a chaotic region in which there is no perceived pattern in the sequence of terms in the series.

Bifurcation corresponds to a situation in which the limiting form of the series is such that $x_n + 2 = x_n$. The graphical analysis can be instructively extended to this case by plotting the result of two iterations $f(x, \lambda)$ for various choices of λ. Here

$$f(x, \lambda) = \lambda x_2(1 - x_2)$$

where

$$x_2 = \lambda x(1 - x)$$

$f(x, \lambda)$ is most conveniently generated on the computer. The results for two choices of λ are shown in Fig. 2 with, as before, the line $y = x$. For $\lambda = 2$, the point of intersection of the two functions corresponds to the single value to which the series tends. The case $\lambda = 3.4$ shows the way in which bifurcation arises. A graphical construction analogous to that deployed in Figs. 1a and 1b converges on one or other of the intersection points A and B, depending on the initial choice of x. It is easily seen that the central intersection point is unstable. The transition to the next bifurcation, in which four values recur cyclically in the original series, arises when, with increasing λ, the modulus of the slope of the $f(x, \lambda)$ curve at the outer points of intersection exceeds unity, as for the single iteration procedure. In this case the iteration tends to one of two rectangular patterns.

Whilst the graphical analysis, which can be extended to cases of repetition after more iterations, provides a valuable insight into the way in which the various iterative regimes arise, the computer-generated *bifurcation tree* in Fig. 3 reveals the full complexity of the iteration possibilities. To generate this diagram, for each choice of λ starting from an arbitrary initial value, say $x = 0.5$, many

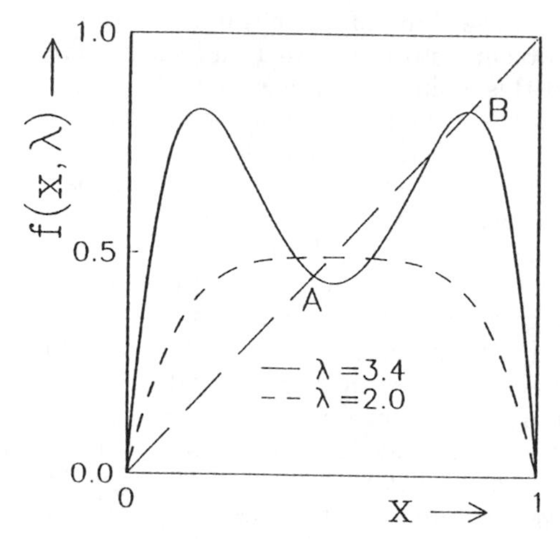

FIG.2 Graphical solution of $x_{n+2} = x_n$ in the iteration of the logistic equation. The solutions are the points of intersection of the 45° line with the curves for $f(x, \lambda)$

terms in the series given by (1) are determined. After discarding the first part of the series to allow for stabilisation to occur, if it will, successive values are plotted on the vertical line representing the choice of λ. For $\lambda < 3$, the one limiting value of the series is repeatedly plotted. When bifurcation arises, two values are printed alternately giving rise to a branching of the tree, with further branching as λ increases. The chaos region, in which whole areas of the diagram are shaded by the printing of dots representing successive values in the iteration, is reached via an infinite series of bifurcations.

This process can be characterized in terms of the values of the ratio

$$\delta_n = \frac{\lambda_n - \lambda_{n-1}}{\lambda_{n+1} - \lambda_n} \cdots \qquad (2)$$

where, λ_{n-1}, λ_n and λ_{n+1} are the values of the control parameter at the $(n-1)$th, nth, and $(n+1)$th bifurcation points. Feigenbaum[5] showed that, for the logistic equation and all other quadratic iterations, the ratio δ_n tends to 4.669 as $n \to \infty$. This *Feigenbaum number* is thus of considerable universality.

The *self-similar* detail of the infinite series of bifurcation steps is contained within a quite restricted range of values of the control parameter λ. In the limit when chaos is reached, it is not without

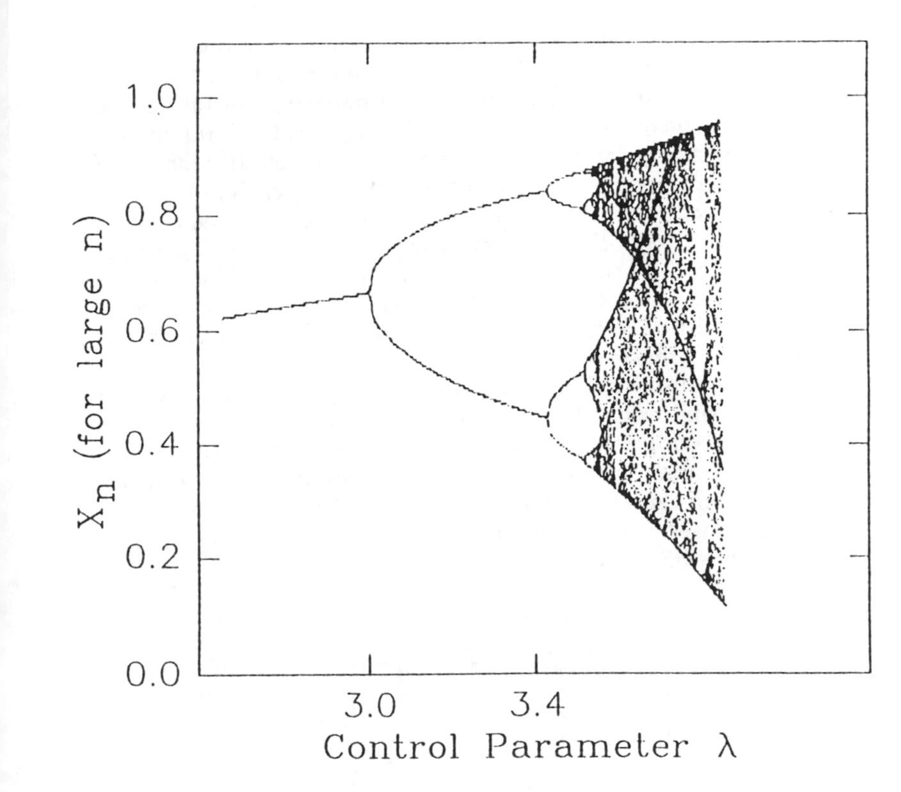

FIG.3 Computer generated bifurcation diagram for the logistic equation

structure. With reference to our experimental work, it is of interest to note the narrow region of *trifurcation* embedded in the chaos region.

Experimental observation of chaos

The observation of chaos in a nonlinear LCR circuit was first reported in papers in 1981[6] and 1982.[7] We are indebted to Dr. Tritton of Newcastle University for initially stimulating our interest in chaos in this type of circuit,[8] and for his subsequent helpful advice. The experimental arrangement is extremely simple. It consists of an inductor, a resistor, and a solid-state diode (serving as a varactor diode) connected in series across the output of a sine wave generator. Because both the effective capacitance and resistance of the diode depend on the polarity *and* the amplitude of the voltage across the device, the current in the circuit does not simply scale with the amplitude of the input a.c. voltage. The response of the LRVD circuit is much more complicated.

At small amplitudes the current is simply a phase-shifted replica of the input sine wave. With increasing amplitude, however, the current may contain the various subharmonics of the drive frequency that correspond to successive bifurcations, or instabilities in phase, which represent chaotic behavior. These changes in the current response may also be brought about by changing the frequency at constant drive voltage amplitude.

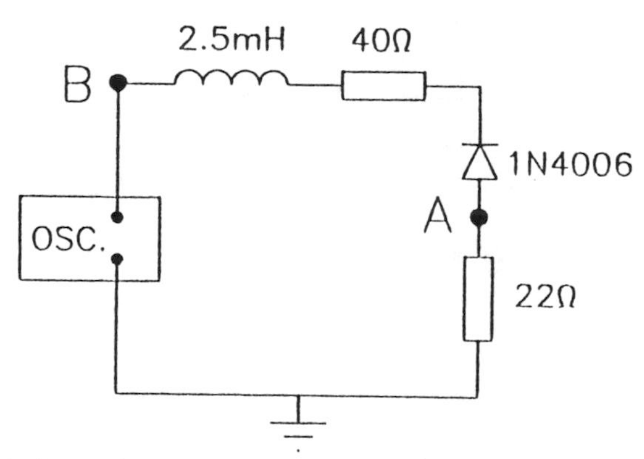

Fig.4 The inductor/varactor diode circuit (LRVD circuit). For x-y display, the x and y inputs to the oscilloscope are connected to points A and B respectively

The effects are most simply observed by displaying the current and input voltage as a Lissajous figure on an oscilloscope screen; Fig. 4 shows a suitable arrangement with a typical choice of components. The current is observed in terms of the voltage it generates across the series resistance. Figure 5 shows some of the effects that arise as the frequency of the input voltage (plotted horizontally) is increased at fixed amplitude. In Fig. 5a, at 100 kHz the current and voltage are simply related; in Fig. 5b the subharmonic in the current waveform gives rise to the double loop pattern corresponding to bifurcation; the case of two bifurcations, in which the current has a quarter-frequency component, is seen in Fig. 5c; and in Fig. 5d the relationship between the current and voltage is chaotic.

The nature of the subharmonic generation is best appreciated by displaying together the voltage and current waveforms on a dual-beam oscilloscope. The cases of bifurcation and chaos are shown in Fig. 6, where it is seen that the mapping of the current on to the voltage waveform is a very complicated matter. Fortunately, for a proper understanding of the phenomena, it is not necessary to record all the details of the current in relation to the drive voltage. It suffices to observe successive values of the current at just one point in the drive frequency cycle. The choice of point is unimportant; we have chosen to observe the current at the zero-crossing of the drive voltage as it goes from negative to positive. We find it simplest to model on the computer the map generated by changing the drive voltage amplitude at constant frequency. Depending on the regime, the observation yields a single current value, two values where there is bifurcation, and, in the limit, a continuum of values in the region of chaos. These values, when plotted as a function of the drive amplitude (or the frequency), generate a *Poincaré map.*

In principle, by using a fast ADC to sample the current synchronously at the voltage crossing point, such a map could be generated instrumentally. This is a considerable project in itself, which we hope will be undertaken by other students in the near future. A much quicker result is obtained by making careful measurements visually on the screen of a dual-beam oscilloscope. We have found it convenient to trigger the scope from the drive signal and, by using the x-shift control, to align the voltage crossing points of interest on central vertical fiducial line on the screen to facilitate the measurement of the corresponding current values. It is not necessary to obtain absolute current values. As long as the y-shift control for the current waveform is not disturbed, the vertical

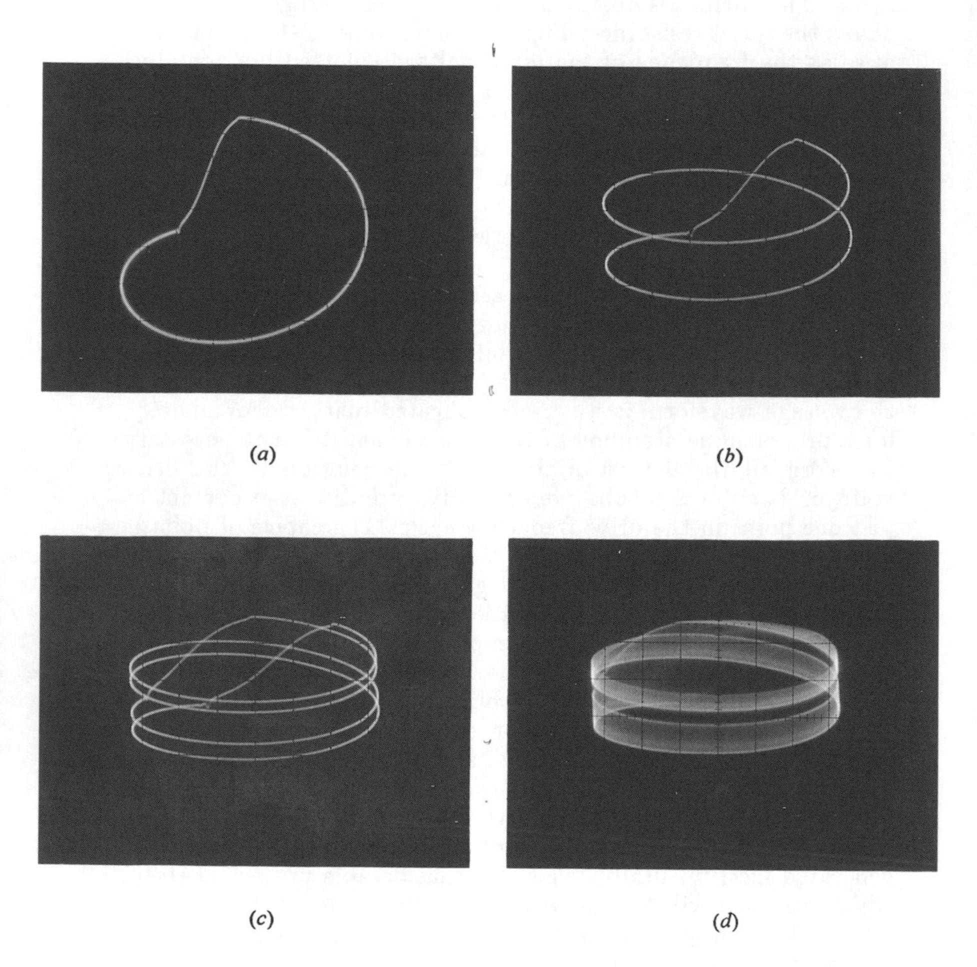

FIG.5 Some effects of changing the frequency of the input to the LRVD circuit of Fig. 4. In the Lissajous figures, x and y represent input voltage and current respectively. Here the input ampitude is kept constant at 6.5 volts pk-to-pk; the frequencies are (a) 100 kHz, (b) 175 kHz, (c) 220 kHz, and (d) 250 kHz

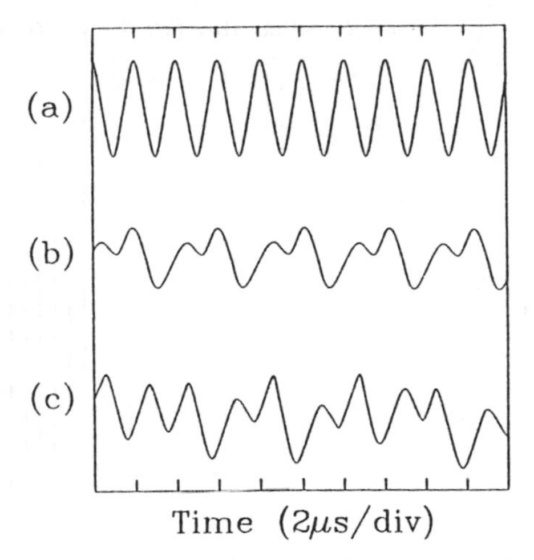

FIG.6 Voltage/current relations at 500 kHz in the LRVD circuit: (a) is the input voltage, and (b) and (c) are current waveforms for voltage inputs of 2v and 3.5v peak-to-peak respectively. The vertical scales are arbitrary - our purpose is only to display the differences in time variation

position of the waveform on the central line generates a perfectly satisfactory map.

Figure 7 shows such a map, generated at 500 kHz. With its bifurcations and regions of chaos, it has the same elements as are found in the logistic equation diagram of Fig. 3; there is even a region of trifurcation where the original frequency is divided by three. Not suprisingly, however, the overall shape of this bifurcation tree is not the same as that for the logistic equation. Moreover, the precise form of the tree is frequency dependent; for a complete representation of the phenomena, a three-dimensional map would be required with frequency as the third axis. What is striking is that the same fundamental features of repeated bifurcation and chaos are found at a wide range of frequencies.

On the left of Fig. 7 a succession of three bifurcations is seen, but the precision of the data that can be obtained directly from the oscilloscope does not permit one to follow the series of divisions very far. It is not possible to establish with any precision δ_∞, the Feigenbaum number for this process, but the ratios here for δ_2 and δ_3 using (2), are 5 and 4, respectively. These are not inconsistent with the limiting value of 4.66.

In its detailed form the bifurcation diagram of Fig. 7 is very

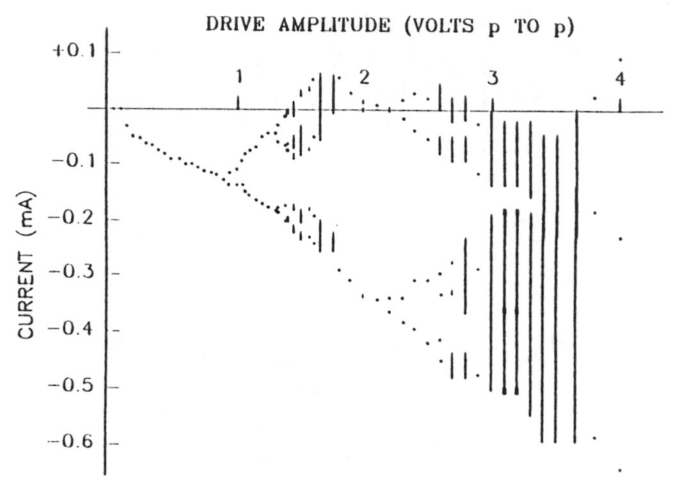

FIG.7 Experimental bifurcation diagram for the LRVD circuit of Fig. 4 at 500 kHz. Points represent observed single values of the current at the input voltage crossing points, as explained in the text. Where bifurcations occur, more than one point is plotted at the given drive amplitude. In the limit of chaos, the band of current values observed is represented by a vertical line

similar to that which can be generated by the computer modeling of the circuit, to which we now turn.

Computer simulation

It has proved to be of considerable pedagogic interest and value to model the response of the inductor/varactor diode circuit on the computer. Figure 8 shows a computer simulation at a frequency of 500 kHz in which the peak-to-peak applied voltage across the circuit, v_{max}, varies from 0 to 6 V. The program was written in FAST BASIC for the Atari ST (see Appendix), and to maintain accuracy, it uses 250 time steps per r.f. cycle. It runs several million times slower than real time; Fig. 8 took some 48 hours to compute!

At each new value of the applied r.f. voltage, a settling time of 20 r.f. periods is allowed. After that, in each cycle, as in the experiment, as the instantaneous applied voltage, V, passes through zero in the positive direction the corresponding current, I, is plotted. Frequency sub-division or chaos are present if I is multivalued.

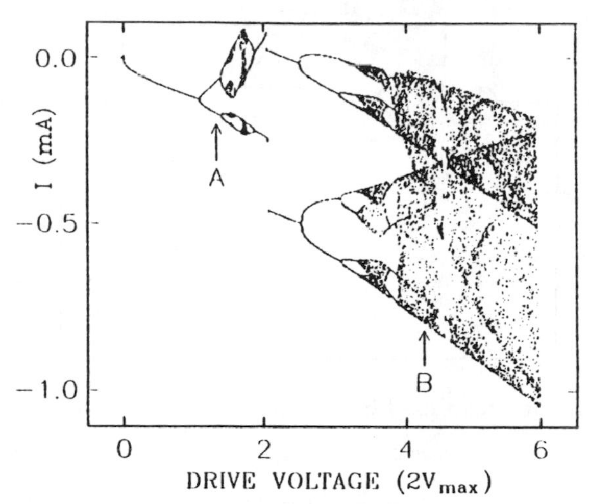

FIG.8 Computer simulation of the bifurcation diagram for the LRVD circuit at an assumed frequency of 500 kHz. Bifurcation at A and chaos at B are represented in Figs. 10a and 10b

For this calculation the varactor diode is represented by a resistance of $10^5\,\Omega$ (conductance $G = 10^{-5}$) in parallel with an ideal (silicon) diode with the threshold voltage $V_0 = 0.6\,V$. When conducting, the current through the diode is assumed to be proportional to $(V_d - V_0)^3$ where V_d is the voltage across the diode. The diode capacitance at zero voltage, C_0, is assumed to be 33 pF and, as explained below, in the model the capacitance decreases slowly from C_0 with reverse bias but increases rapidly with bias in the forward direction. The inductance value chosen is 2.5 mH, and the total series resistance of the circuit is set at $60\,\Omega$.

The iteration algorithm can be understood by reference to Fig. 9. Let the instantaneous voltage applied to the circuit be described by the equation

$$V = V_{\mathrm{max}} \cos(\omega t) \tag{3}$$

The current I flows partly through the diode and partly through the capacitor so that $I = I_d + I_c$. Thus, in Fig. 9 we have

$$V_d = V - L\frac{dI}{dt} - RI. \tag{4}$$

Knowing V_d and the parameters of the model diode allows the diode current I_d and the diode capacitance C to be calculated.

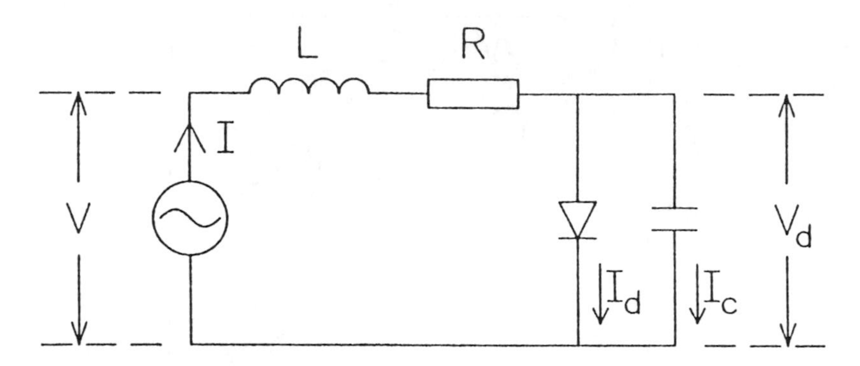

FIG.9 Equivalent circuit used in the computer simulation

From these

$$I_c = I - I_d \tag{5}$$

and

$$\frac{dV_d}{dt} = \frac{I_c}{C} \tag{6}$$

can be computed.

Differentiating (4) and using (5) and (6), we have

$$\frac{I - I_d}{C} = \frac{dV}{dt} - L\frac{d^2I}{dt^2} - R\frac{dI}{dt} \tag{7}$$

In the iteration, let I_2 be the value of I at time T, I_1 the previous value at time $T - \Delta T$, and I_3 the next value at time $T + \Delta T$. Then

$$\frac{dI}{dt} \approx \frac{(I_3 - I_1)}{2\Delta T} \tag{8}$$

and

$$\frac{d^2I}{dt^2} \approx \frac{I_3 - 2I_2 + I_1}{\Delta T^2} \tag{9}$$

Differentiating (3) and using (8) and (9) in equation (7) gives, after some rearrangement,

$$I_3 = (1+F)^{-1}\left\{2I_2 - I_1 - E\left[\omega V_{\max}\sin(\omega t) + \frac{(I - I_d)}{C}\right] + FI_1\right\} \tag{10}$$

where $E = \Delta T^2/L$ and $F = R\Delta T/2L$. The complication of the voltage dependence of the capacitance of the diode is handled some seven lines earlier in the program. For positive values of V_d, we take

$$C(V_d) = C_0(1 + 100V_d^2)$$

For negative bias, we use the form

$$C(V_d) = \frac{0.89C_0}{(0.8 - V_d)^{1/2}}$$

Note that the program ensures that the value of the current is plotted as V passes through zero in the positive direction. If this is not done, two bifurcation diagrams will be produced together.

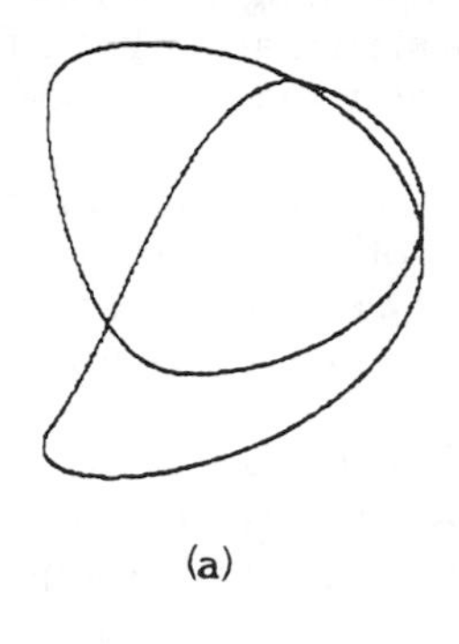

(a)

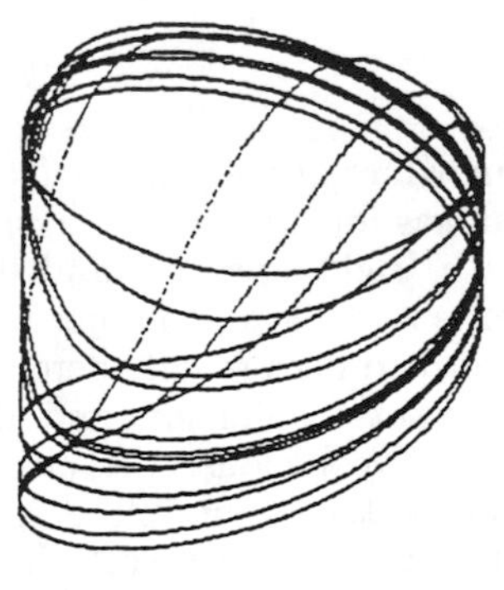

(b)

FIG.10 Current/voltage loci in the computer simulation. Current and voltage are plotted on the vertical and horizontal axes respectively. (a) bifurcation at A in Fig. 8; (b) chaos at B in Fig. 8

In addition, while the program is running, the frequency and voltage can be altered using appropriate keys, and individual current-voltage characteristics can be plotted as Lissajous figures. An example is shown in Fig. 10a in which frequency halving occurs at a drive frequency of 500 kHz; this corresponds to position A in Fig. 8. Figure 10b shows cycles of chaotic behavior corresponding to position B in Fig. 8.

The overall pattern of the computed bifurcation diagram of Fig. 8 parallels quite closely the experimental observations presented in Fig. 7. In both cases, two complex regions of repeated bifurcation and chaos are separated by a bridging range of voltage in which the two alternating current values change only slowly. In Figs. 7 and 8 the first bifurcation occurs at peak voltages of 0.47 and 0.55 V, respectively. The corresponding values for the onset of the second cascade of bifurcations are 1.1 and 1.27 V. Given the limitations of the model, the agreement in the voltage scales for the two diagrams is surprisingly good.

As might be expected, in the modelling, the regions of chaos and the detailed shapes of both the bifurcation plots and the current-voltage loci depend very sensitively upon the model parameters used. It is clear that the essential requirements have been incorporated in the model, but it would be interesting to investigate the range of parametrisation that gives rise to chaos. Unfortunately, this is not really practicable without employing a much faster computer program and a more powerful computer.

Conclusion

Chaos is a subject in which we are invited to 'second-guess' nature, and its study makes for interest in physics courses which are mainly concerned with 'right answers' and do not usually allow much scope for students to offer an opinion. The inductor/varactor diode experiment described here readily demonstrates many of the essential features of chaotic behavior, and it gives considerable scope for student initiative because the circuit behavior is governed by two parameters, the input voltage and the frequency. It has the further advantage for project work that, with patience, the circuit response can be rather well modelled using a fairly small computer. The interplay of experimental observation and modelling illustrates very clearly the essential nature of the scientific method.

References

1 For an early important review see: R.M. May, *Nature*, **261**, 459 (1976)
2 For a nontechnical introduction to chaos, see: J. Gleick, *Chaos: making a new science*, Heinemann (1988)
3 A comprehensive discussion of chaos and related topics will be found in: A.B. Pippard, *Response and Stability*, Cambridge (1985)
4 The dripping tap problem is included in the following accessible review: J. Crutchfield, J. Farmer, N. Packard, and R. Shaw, *Sci. Am.* **255**, 38 (1986)
5 M.J. Feigenbaum, *Phys. Lett*, **74a**, 375 (1979)
6 P.S. Linsay, *Phys. Rev. Lett.*, **47**, 1349 (1981)
7 J. Testa, J. Perez, and C. Jeffries, *Phys. Rev. Lett*, **48**, 714 (1982)
8 D. Tritton, *New Scientist*, July (1986)

Appendix

(See Appendix on page 208.)

Computer analysis of a water rocket

MEMBERS OF THE SIXTH FORM
Wolverhampton Grammar School, Wolverhampton, UK

Abstract: A number of physical principles is illustrated by the familiar water rocket. A computer program is described that calculates the vertical height reached by a rocket, using a simple form of iterative analysis.

The water rocket is an ideal object for study by the Sixth Form. It consists of a bottle-shaped chamber, partially filled with a propellant (e.g., water). The chamber is inverted and pumped up with compressed air. When launched, the propellant issues from the nozzle and the rocket is thrust upward. The physical principles by which it operates are fundamentally simple, and yet there are sufficient unknowns in the system to make some of the results interesting, even surprising. In addition, it is possible to show how practical and theoretical studies must go hand in hand if progress is to be made.

The computer program is shown in Fig. 1 and is written in BBC BASIC. It requests the initial starting parameters concerning the rocket and its propellant, and then procedes to calculate the maximum height reached by the rocket. Some of the variables may need explaining.

In line 40, K is what we called the loss rate constant, and determines the rate at which propellant is ejected from the nozzle of the rocket. Assuming that the area of the nozzle is much smaller than the cross section area of the rocket itself and that the flow is not too turbulent, a simple application of Bernoulli's principle shows that the rate at which fluid is ejected is given by

$$dM/dt = KA\sqrt{(}P_1 - P_0)$$

where A is the area of the nozzle, $(P_1 - P_0)$ is the pressure difference and K is a constant equal to $\sqrt{2} \times$ density. For water, K therefore has the theoretical value of 44.7 $\mathrm{kg}^{1/2}\mathrm{m}^{-3/2}$.

Since this constant plays a vital role in the program (it appears in line 240), it was important to check this value experimentally. This was done by fixing a nozzle to the bottom of a tall drainpipe

```
 10 REM WGS WATER ROCKET PROJECT 1987
 20 REM Simple program
 30
 40 K=44.7:G=9.81:D=1000:AP=100000
 50
 60 REPEAT
 70 PRINT"Input the following in SI units"'
 80   INPUT"Mass of rocket     ";MR
 90   INPUT"Volume of rocket   ";VR
100   INPUT"Area of nozzle     ";AREA
110   INPUT"Mass of fuel       ";MF
120   INPUT"Absolute pressure  ";PO
130   INPUT"Drag coefficient   ";DR
140
150   VAO=VR-(MF/D)
160   LR=0:F=0:V=0:A=0:H=0
170   T=0.002:ET=0
180   Hmax=0:Vmax=0:Amax=0
190
200   REPEAT
210     IF MF=O THEN T=0.1
220     VA=VR-(MF/D)
230     P=(PO*VAO)/VA
240     LR=SQR(P-AP)*AREA*K
250     MF=MF-(LR*T)
260     IF MF<=O THEN MF=0:LR=0
270     F=LR*LR/(AREA*D)-DR*V*V*SGN(V)
280     A=F/(MR+MF)-G
290     V=V+A*T
300     H=H+V*T:IF H>Hmax THEN Hmax=H
310     ET=ET+T
320   UNTIL H<Hmax
330
340   PRINT"Max height    = ";Hmax'
350 UNTIL FALSE
```

FIG.1

and measuring the time for various amounts of water to drain out. This experiment was in excellent agreement with the theory yielding a value of $(43 \pm 3)\,\mathrm{kg}^{1/2}\,\mathrm{m}^{-3/2}$.

Returning to line 40, G is, of course, the acceleration due to gravity, D is the density of water, and AP is atmospheric pressure.

Only one of the initial parameters needs any special comment, DR the drag coefficient in line 130. It was assumed that the force of air resistance would be proportional to the square of the velocity. DR is defined therefore as the force of air resistance per unit velocity.[2] Unfortunately, we were unable to make an accurate measurement of this quantity, nor were we able to check the validity of this assumption. In practice, as one might expect, air resistance proved to be a very significant factor and further work will be done in this area.

VAO in line 150 is the original volume of air in the rocket.

T in line 170 is the time increment over which the calculations are made. Initially, it is set to 2 ms. Later, when all the propellant is used up, the accelerations are much smaller and T is set to 100 ms (in line 210). This greatly speeds up the calculations. ET is the elapsed time.

The main loop (lines 200 to 320) then procedes to calculate the new values of all the variables after one time increment T. The sequence of calculations is shown in Fig. 2.

Line 230 uses Boyle's law to calculate the pressure in the rocket. It could well be argued that the expansion of the gas in the rocket is adiabatic rather than isothermal, but as the change in pressure is relatively small, this point was not considered further.

Line 260 checks to see if all the fuel has been used up.

Line 270 calculates the force on the rocket (F) using Newton's second law to calculate the thrust on the rocket and the drag coefficient to calculate the air resistance.

Finally, lines 280 to 300 calculate the acceleration, velocity and vertical height of the rocket.

While this program demonstrates the algorithm employed, it is not the program that we actually used. Two further programs were developed. The first was essentially the same but with a more sophisticated input routine and a graphical output; the second enabled us to plot the maximum height reached as a function of any one of the starting parameters. It was this program that revealed some of the surprises. Two samples are shown in Figs. 3 and 4. The first illustrates the result that the maximum height of the rocket is remarkably insensitive to the mass of propellant used, and the second demonstrates that, for given air resistance, the optimum

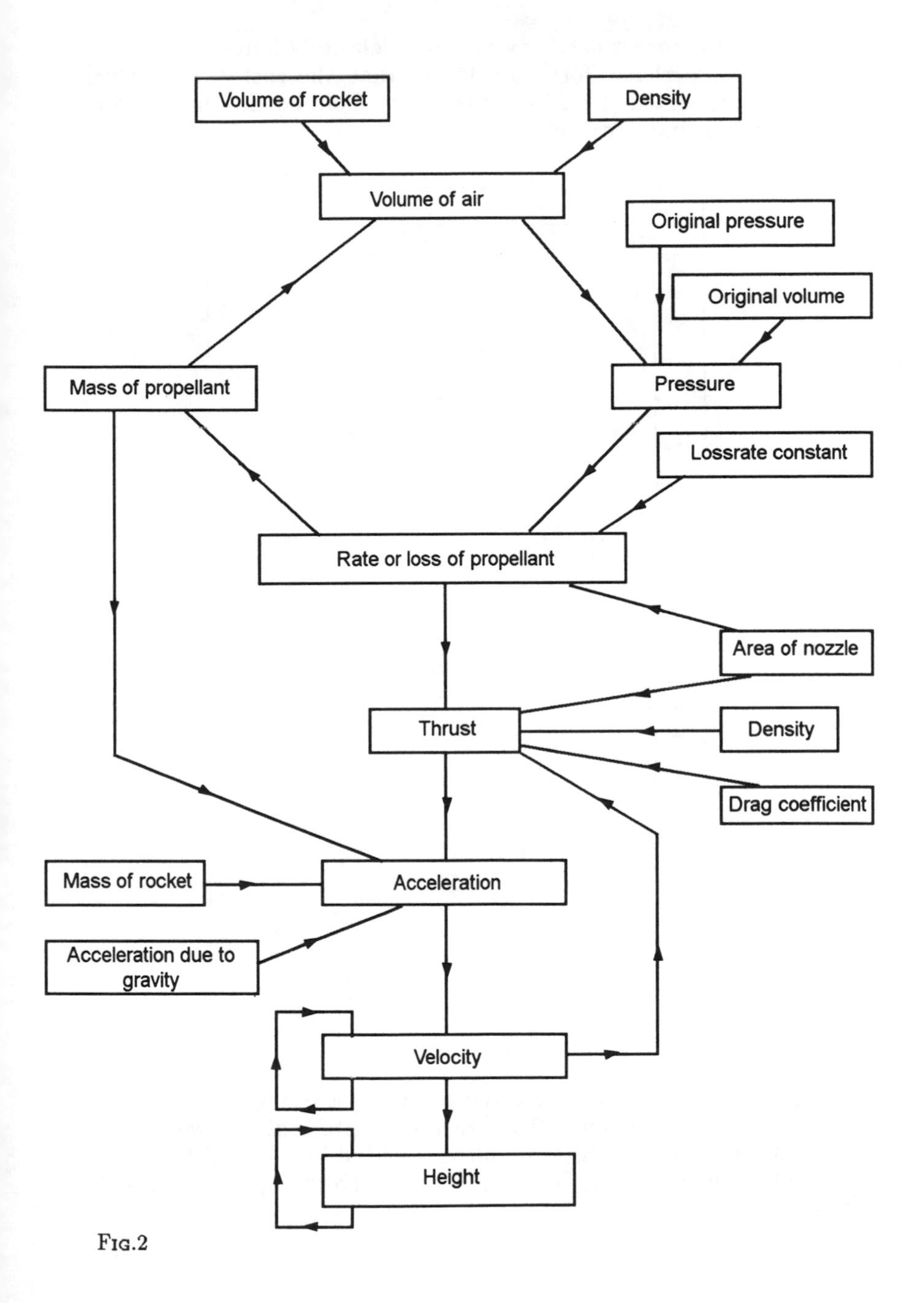
Volume of rocket
Density
Volume of air
Original pressure
Original volume
Mass of propellant
Pressure
Lossrate constant
Rate or loss of propellant
Area of nozzle
Thrust
Density
Drag coefficient
Mass of rocket
Acceleration
Acceleration due to gravity
Velocity
Height

Fig.2

mass of the rocket itself is small, but definitely finite.

It was, perhaps, fortunate for us that the rocket with which we experimented (a plastic 2 litre Coca–Cola bottle) proved to be optimal in many respects!

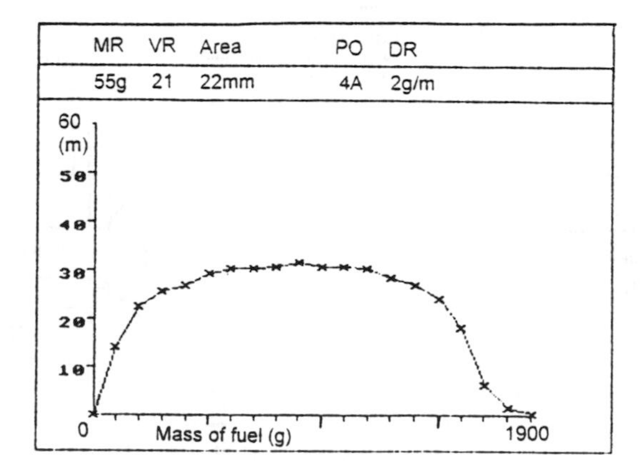

Fig.3

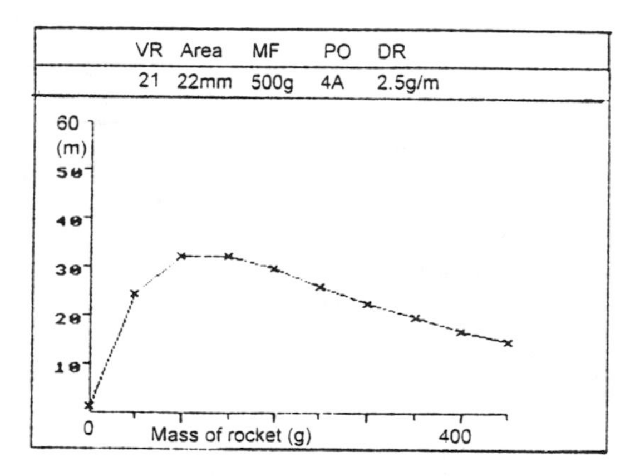

Fig.4

Anyone interested in receiving copies of other computer programs mentioned (for a BBC microcomputer only) is welcome to send a 40 track disk to J.O. Linton, Head of Physics, Wolverhampton Grammar School, Compton Road, Wolverhampton, WV6 9RB, UK.

Gyroscopic motion

G. LANCASTER
Department of Physics, University of Keele, Keele, Staffordshire, ST5 5BG, UK

Abstract: In this first year experiment, a large stainless steel ball (10 cm diameter) is set spinning on a low friction air bearing and the precessional motion due to an applied mechanical torque is studied. Important aspects of the experiment are an investigation of the frictional torque in the air bearing and the technique used to measure the angular velocity of the ball.

(It must be emphasised that for students this is an experiment, not just a demonstration of a physical phenomenon. They have to assess the problem and the apparatus provided and manipulate the theory to some extent and then have to plan their experimental work. A mathematical function has to be found to fit some of their results, some data analysis has to be performed, and importantly, a summary has to be written.)

Introduction

For the rotational motion of a body, the relationship between the applied torque Γ, the moment of inertia I, and the angular velocity ω is

$$\Gamma = I\dot{\omega} \tag{1}$$

Since the angular momentum L of the body is equal to $I\omega$, it follows that

$$\Gamma = \dot{L}$$

i.e., torque = rate of change of angular momentum.

It can be shown (*students are referred to lecture notes/texts*) that, if a torque Γ acts on an otherwise freely rotating body, the axis of rotation precesses with angular velocity $\mathbf{\Omega}$ where

$$\mathbf{\Gamma} = \mathbf{\Omega} \times \mathbf{L} \tag{2}$$

(see Fig. 1) or

$$\Omega = \frac{\Gamma}{L} = \frac{\Gamma}{I\omega} \tag{3}$$

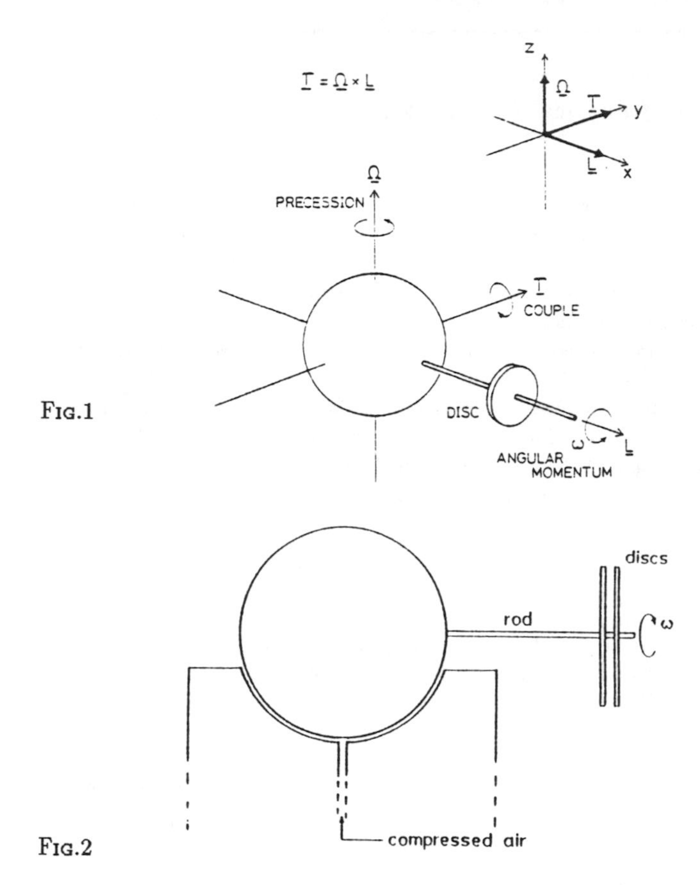

Fig.1

Fig.2

Aims of the experiment

Important aims are to confirm the spatial relationship between $\mathbf{\Gamma}, \mathbf{\Omega}$ and $\mathbf{L}$ as specified by (2) and, using (3), to obtain an experimental value for the moment of inertia of the steel ball, which can be set spinning whilst supported on an air bearing (see Fig. 2).

Students also have to determine whether the frictional torque associated with the air bearing is significant in relation to the magnitudes of the applied torques.

There are two important advantages of using a large steel ball. First, the ball has a large angular momentum for a small angular velocity ω, which means that the ball can be set spinning by hand ($\sim$ 10 rev/sec). Second, the aluminum disks attached to an aluminium

rod, which are used to provide the applied torque (see Fig. 2) have a negligible moment of inertia compared to that of the ball so that the system can be considered simply as a spinning sphere.

Experiment I

To investigate the frictional torque of the air bearing

It is not known *a priori* whether the frictional torque associated with the air bearing is significant compared to the applied torques ($\sim 0.01\,\mathrm{Nm}$).

If the frictional torque is $\Gamma(\omega)$, (1) shows that

$$|\Gamma(\omega)| = I\left|\frac{d\omega}{dt}\right| \tag{4}$$

For a sphere of radius R and mass M

$$I = \frac{2MR^2}{5} \tag{5}$$

For a steel ball with $R = 5.0\,\mathrm{cm}$ and $M = 4.29\,\mathrm{kg}$, we find that $I_{\mathrm{calc}} = 4.1 \times 10^{-3}\,\mathrm{kg\,m^2}$.

An electric drill with a cloth polishing wheel attached is used to set the steel ball spinning at about 3 000 rpm (with the aluminum rod vertical and with no disks attached). The angular velocity of the ball is measured as a function of time (see Fig. 3) and $|d\omega/dt|$ can be determined over a large range of velocities and, in particular, at the low angular velocities used in Experiment II when the precession is being investigated.

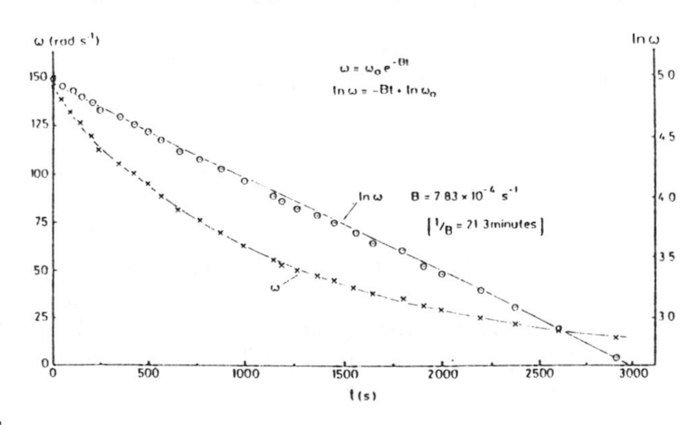

Fig.3

Students are asked to plot a graph of ω versus t and to determine the functional relationship between these quantities. Usually, after inspecting their graph, and some thought, they guess that ω declines exponentially (although $\omega\alpha \propto t^{-n}$ is tried quite often).

Since it is found that

$$\omega = \omega_0 e^{-Bt}$$

it follows that

$$\ln \omega = -Bt + \ln \omega_0$$

The 'decay constant' B can thus be determined from the linear graph of $\ln \omega$ against t (see Fig. 3).

Also,

$$\left| \frac{d\omega}{dt} \right| = B\omega$$

so that $|d\omega/dt|$, and hence the frictional torque can be found over the measured range of values of ω and, in particular, at the small values of ω used in Experiment II.

It transpires that the magnitude of the frictional torque seems to be proportional to angular speed, as indicated by the close fit to the exponential decay.

If the frictional torque were proportional to ω^2, the relation between $1/\omega$ and t should be linear. However, the experimental results do not support this hypothesis.

Measurement of the period of rotation of the steel ball

For some time a stroboscope was used to measure the period of rotation, with a signal generator used to drive the stroboscope. The frequency of the signal generator was determined with a frequency counter.

Although stainless steel is non-magnetic, nominally, the ball was discovered to be weakly magnetised. An easily measurable induced voltage is generated in a pick-up coil placed near to the rotating ball. The voltage waveform displayed on a CRT is complex, but a repetitive period equal to that of the spinning ball can be readily identified and measured using the calibrated time-base of the CRT. The complexity of the waveform reflects, no doubt, the non-uniform magnetization of the steel ball.

Experiment II

To verify (2) and to determine the moment of inertia of the ball

Students should discover the range of torques that can be obtained using the disks either singly, or in combination, at various positions along the aluminum rod and plan their subsequent measurements accordingly.

By rearranging the *scalar* form of (3) it can be shown that

$$\Omega\omega = \frac{\Gamma + \Gamma_0}{I} \tag{6}$$

where Γ_0 is the torque due to the hole in the steel ball (which accommodates the aluminum rod) and also due to the rod itself.

After measuring the precessional angular velocity as a function of Γ and ω students plot a graph of $\Omega\omega$ vs Γ (see Fig. 4). The linearity of the graph is evidence in support of the theory and a value for I can be determined from the slope of the graph (agreement to within about 1% of the calculated value can be obtained usually).

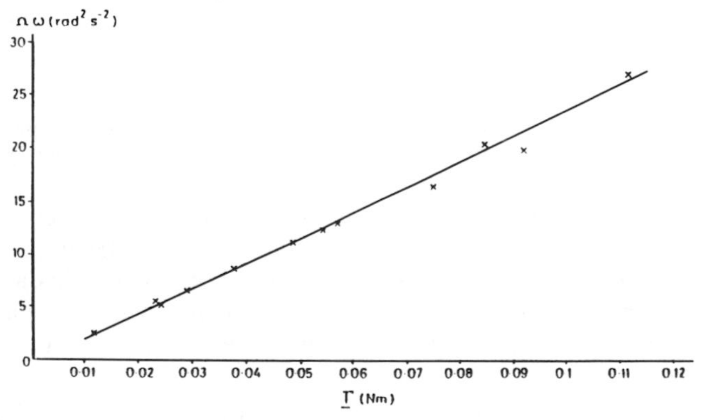

FIG.4

The intercept of the graph is usually very close to zero, within experimental error, which indicates that, coincidentally, the aluminium rod almost exactly compensates for the missing steel from the hole. The position of the centre of gravity of the ball with or without the aluminium rod inserted into the hole can be determined from measurements of the period of oscillation of the ball when it is not spinning.

Stokes viscometer and other microcomputer applications

A.W. REED

Physics Division, Department of Applied Sciences, Staffordshire Polytechnic, Stoke-on-Trent, ST4 2DE, UK

Abstract: Three applications of microcomputers in Physics education are described. The major one is an adaptation of the Stokes' method for viscosity to introduce students to the use of microcomputers in measurement systems.

Microcomputer – based viscometer

Stokes' method of determining viscosity of, say, glycerol is not the most obvious use of a microcomputer on-line to an experiment, since the measurement rate is neither too fast nor inconveniently slow for the experimenter. However, the system described originated as the first part of a student project, where simple BASIC programming could produce a positive result.

It was completed at a time when it was decided to computerize a number of standard physics experiments. Still in its original relatively crude form, the apparatus has survived over two years of use by several first year classes.

A glass tube, 0.5 m × 25 mm in diameter and filled with glycerol, is clipped to a vertically mounted wooden frame. Circuit boards are fixed either side of the tube, about half-way down. This allows three light activated switches (LAS) with a 50 mm vertical spacing to be illuminated through the tube by three 5-mm high-intensity red LEDs.

Ball bearings (4 mm in diameter) are dropped through a narrow guide tube and fall along the axis of the tube, casting shadows on each LAS in turn.

The LAS outputs are fed to three of the user port data lines of a BBC microcomputer. The function of the computer is to measure the two 50-mm falltimes, check that they are closely similar, and produce a printout of mean terminal velocities. The student takes all other measurements, except the tube bore (to avoid disturbing guide-tube alignment). The computer will only release a mean viscosity value if the student enters a similar one.

The software has been designed to be particularly user friendly. All instructions are given on screen, using color to aid presentation and sound to signal warnings. Input errors and nonterminal velocities are allowed for, and a run is aborted if detection fails.

By making the standard corrections for finite dimensions, quite good results are obtained.

There is often a tendency for terminal velocity to increase during an experiment. Two possible reasons are:

(i) the balls drawing down surface absorbed water;

(ii) the balls introducing air bubbles.

The second seems more likely, as the tube is normally kept sealed, and allowing a 'settling time' between balls reduces the effect. The system can be affected by strong direct sunlight.

The complete experiment can be repeated several times before the ball bearings need to be retrieved, although this can be done cleanly with a strong magnet and paper tissue.

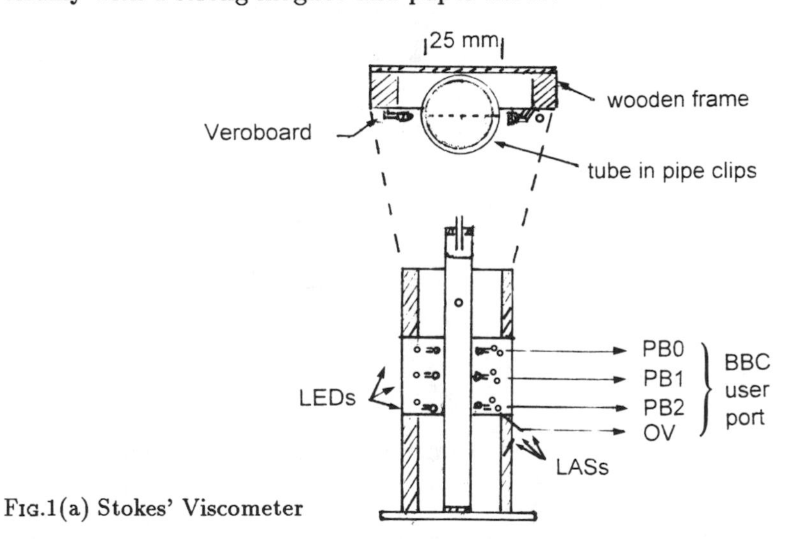

FIG.1(a) Stokes' Viscometer

Microcomputer-based data logger

It is considered important that science students should graduate as 'intelligent users' of microcomputer based instrumentation, i.e., that they should have the ability to design a simple interface and generate the relevant software.

The following experiment has been in use for several years as an introduction to interfacing.

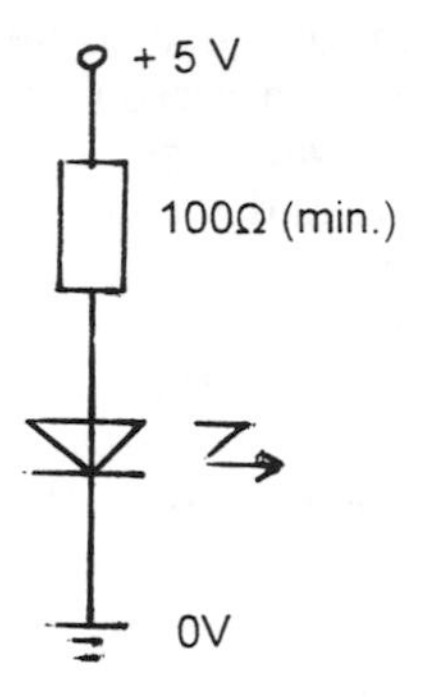

FIG.1(b) LED

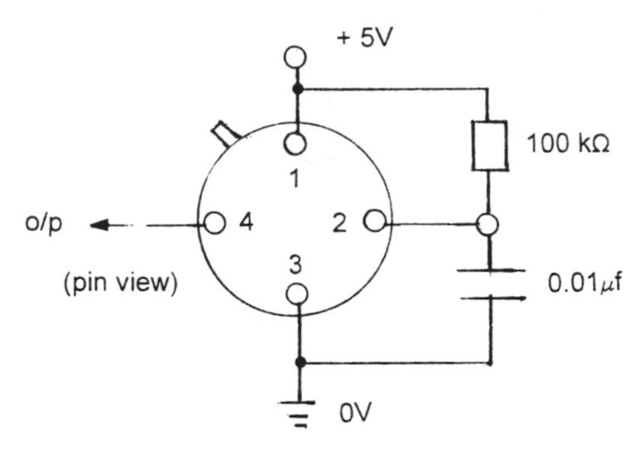

FIG.1(c) LAS

Students, with a limited amount of BASIC programming experience, are provided with a microcomputer and an analog-to-digital converter module (ADC) already interconnected. The labsheet gives brief details of microcomputer operation, the function of the user port, and ADC action, together with details of how to address the various user port registers and access the timer.

The students are first asked to generate a sampling sub-routine, and calibrate the ADC against known input voltages. They are then asked to generate a program to produce a table of input voltages and sample times, with a pre-selected time interval, i.e., to produce a *data logger*.

Further exercises are to generate programs to:-

(i) sample continuously, and plot number of occurrences against input voltages, i.e., simulate a multichannel analyzer;
(ii) sample and plot sinusoidal waveforms of various frequencies (ADC in bipolar mode). This is particularly useful in illustrating sampling theory and aliasing.

Providing they have previously gained an adequate background in programming in the appropriate language, students find the experiment straightforward and satisfying.

An ADC module can be made simply cheaply ($ 10) using a type 448 in the data sheet minimum circuit.

Simple CAD

As part of their Introductory Electronics course, students undertake a number of design, build and test exercises. They are encouraged to use a suite of associated programs to *check* their design calculations, partly as an introduction to the CAD, which they meet later in their course.

Again the software was written with ease of use and pleasing presentation in mind. Circuit diagrams are given on screen along with optimum preferred component values and predicted performance figures.

(Further details of the Stokes' Viscometer and a copy of the program may be obtained by supplying the author with a floppy disk).

Summary of typical results

Ball diameters in the range 3.97——4 mm gave terminal velocities in the range $0.0546 - -0.091\,\mathrm{m\,s^{-1}}$. The mean velocity of glycerol was found to be $0.789\,\mathrm{N\,cm^{-2}}$ with a standard deviation of 0.94.

Vibrating bar depth gauge

P.J. McDonald and E.P. O'Reily*

The first year laboratory course at the University of Surrey is significantly different from most first year laboratories based on traditional one-day set experiments. Our scheme consists of two basic elements. In the first few weeks the students perform very simple experiments lasting from ten to sixty minutes designed to familiarise them with basic equipment and experimental methods. From then on they do relatively open ended mini-projects and activities each lasting five whole days. All the projects (of which the students do three or four from about twenty) are designed to support lecture courses in physics and to teach several practical skills in parallel, including experimental design and construction and data analysis. A few more difficult projects include physics, which is new to the students. Most of the projects include an electronics element and some include microcomputing (elementary hardware and software). As part of the program all the students perform an activity devoted to electronics.

An important aspect of the scheme is that the projects are supported by only a minimal set of notes designed to stimulate and guide rather than instruct. These notes have included reprints of published physics research and component data sheets, for example. Guidance is also available from demonstrators. Assessment is primarily by means of written reports.

We have adopted this approach because it was felt that traditional first year laboratory classes, involving short set experiments, developed technique but did not sufficiently encourage investigative and analytic skills. Since most of our undergraduates spend their third year in professional or industrial training they require these skills early on. An extra bonus has been the added interest of both staff and students in the laboratory classes.

The *Depth Gauge* is a typical mini-project of medium difficulty designed to take five whole days in the laboratory. It was developed out of an industrial research programme. The students are presented with a vertically supported Perspex bar about 1 m long. Experience suggests that a metre rule is not suitable. We use a bar with a cross section of $2 \times 20\,\text{mm}^2$. The bar is made to vibrate

* Reprinted with permission from Physics Education

by a mechanical vibrator attached to the upper end driven by a signal generator. For our bar the frequencies of interest are of the order of 1 − −50 Hz. The amplitude of vibration of the bar is detected using a small permanent magnet attached to the bar which moves relative to a fixed solenoid (Fig. 1) 1. The students have to realise that the signal from the solenoid needs amplification, and then have to build a simple amplifier circuit typically using a 741 operational amplifier with a gain of about 10. They also have to realise that the positioning of the magnet is fairly critical and depends on the mode being observed. The detected signal can then be displayed on an osciloscope (preferably with storage or screen persistence as the frequencies of vibration are low).

The guidance notes as given to the students suggest the following ideas.

(1) Study the motion of the bar and observe a few resonances. Find an empirical relationship between the resonant frequency, the length of the bar, and the vibration mode number. Show how the shape of the first few characteristic functions differ from those of a vibrating string.

(2) The motion of the bar is further damped if the bar is placed in a liquid. Develop the system into a useful depth gauge for water up to 60 cm *deep.*

(3) Likewise, develop the system into a means of measuring viscosity. A commercial viscometer is available in the laboratory and can be used to calibrate your device.

For the depth gauge water damps the end of the bar changing the mode frequencies. Our system works best for the third or fourth mode. Lower modes are hard to track whilst higher modes suffer from overlap in the depth range 0 to 60 cm. It is then possible to calibrate the system to within 1 cm over substantial parts of the designated depth range. Typical results for the second mode of our system are shown in Fig. 2. It is seen that in this case the mode frequency is sensitive to depth variations over most of the range. At depths of around 30 cm, however, the water level is such as to be near a node where there is little bar motion. Consequently sensitivity is lost in this region for the 2nd node.

As a viscometer the apparatus as described has been made to distinguish between water and glycerine only. The liquid tends to damp the end of the bar so that for most viscous liquids a node is generated at or close to the liquid surface. For a bar of our dimensions the motion is then dominated by that part of the bar in air making frequency discrimination between different liquids poor. However, experience suggests that students will find ways to

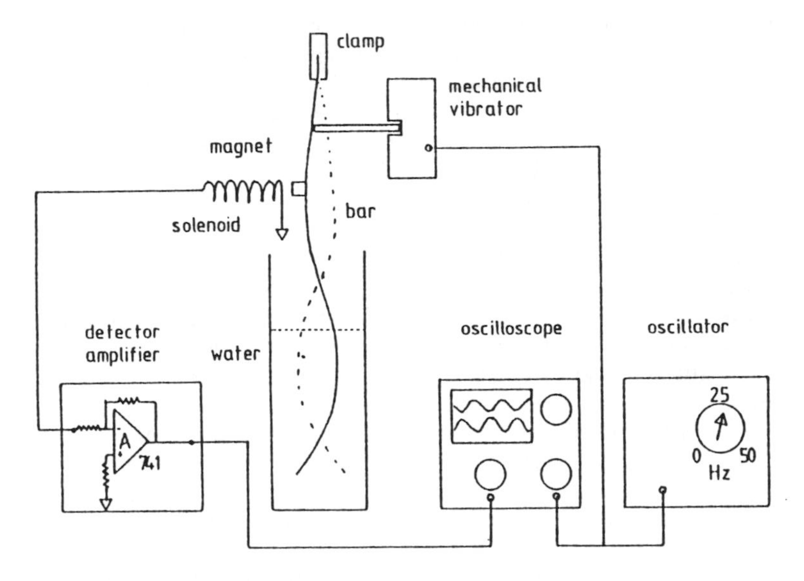

FIG.1 The experimental setup

improve sensitivity. Theoretically this is not difficult for a system of other dimensions in which the bar is more fully immersed in the test liquid.

Students are not required to understand the detailed mathematics of the system. They are, however, introduced to the equation of motion for a rigid bar[1]

$$\frac{\partial^4 y}{\partial x^4} = -\frac{\rho}{Q\kappa^2}\frac{\partial^2 y}{\partial t^2} \tag{1}$$

where the rod lies along Ox and is in motion along Oy, ρ is the mass per unit length of the bar, Q is Young's modulus, and κ is the radius of gyration (a simple function of the cross sectional dimensions). The students are encouraged to see if their results for (1) agree with theoretical studies, which predict the approximate relation

$$\nu_n = K/L^2(n - \frac{1}{2})^2$$

where ν_n is the n-th resonant frequency, L is the bar length, and K is a constant that depends on the bar geometry and material.

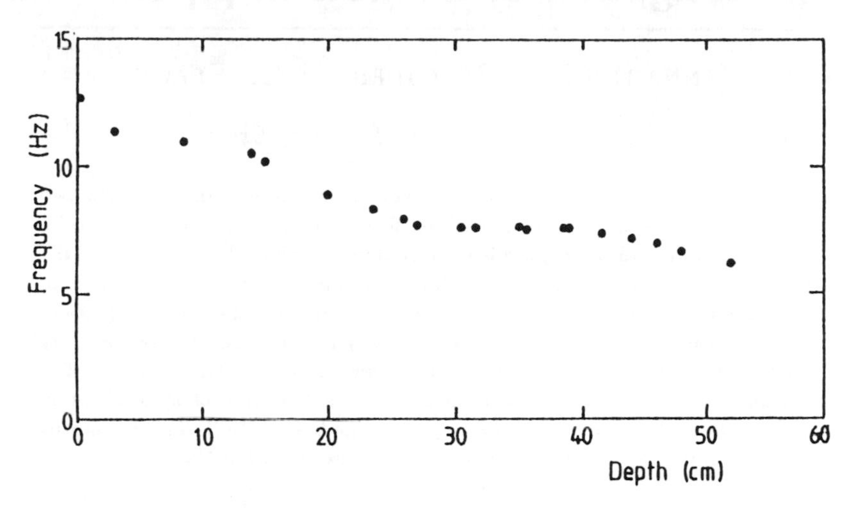

FIG.2 Sample results for the second mode of our system

A report is required at the end of the project. It is expected to include an analysis of the results obtained using the device(s) and a discussion of their reliability accuracy and precision. A discussion of the amplifier is also required.

References

1 P.M. Morse, Vibration and Sound, Chapter 4

Investigation of air flow in pipes

W. Vennart, I.R. Summers, F.C. Flack, and R.E. Ellis
Department of Physics, University of Exeter, Exeter, UK

Abstract: A simple experiment enabling the measurement of the two regimes (laminar and turbulent) of air flow through a plastic (PVC) tube is described. The air flow speed is measured using volume flow meters and the pressure drop along the tube with fluid manometers. A logarithmic plot of pressure drop versus volume flow rate yields a compound straight line graph in which the low speed region (laminar flow) has a slope of approximately one and the high speed region (turbulent flow) a slope of 1.85. There is also a transition region. This transition starts at a flow rate of approximately $1.7 \times 10^{-4}\,\mathrm{m^3 s^{-1}}$ for which the Reynolds number is 2 060. This apparatus provides data that are reproducible and accurate to within 5%.

A fluid can flow according to one of two flow regimes, namely, laminar or turbulent. The former, observed at low speeds, is characterised by smooth flow of the fluid along well-defined flow lines and the latter, at high speeds, by large random fluctuations superimposed on the basic flow pattern. Most laboratory experiments involving fluid flow use water primarily to monitor and quantify the laminar flow regime only.

In this experiment both flow regimes are measured for air flowing through a plastic (PVC) tube of internal diameter 6.0 mm and length 500 mm (Fig. 1). This straight section of tube is connected to a variable compressed air supply via two Rotameter volume flow-meters. An extra 2.5 m continuous length extension of the tube is included between the flow-meters and the measurement section so that any turbulence introduced by the flow-meters is damped out before the air reaches the 500 mm measurement section. Two small holes of diameter 1 mm are drilled into the tube so that manometers can be connected (using PVC tubing) to measure pressure drop. The whole apparatus is constructed so that any discontinuities in the air-flow are kept to a minimum. As a useful addition, a small microphone can be sealed into the exit end of the pipe to monitor the onset of turbulent flow, since sound is only generated when turbulent flow is present.

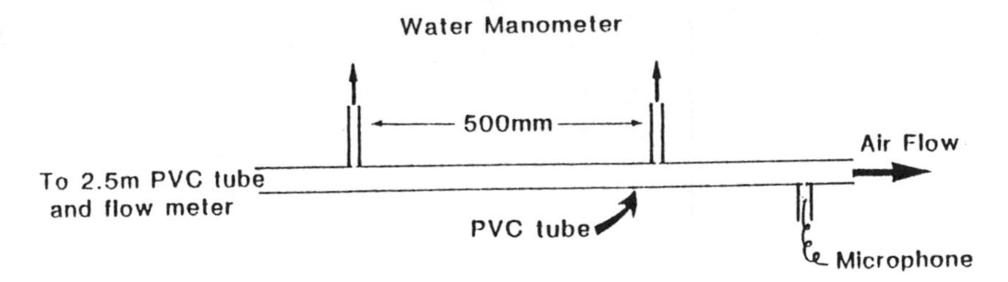

FIG.1 Schematic diagram of the apparatus

At low speeds the relationship between the pressure drop ΔP and volume flow rate Q is given by the Poiseuille's relation

$$\Delta P = \frac{128\mu L}{\pi d^4} Q \tag{1}$$

where μ is the viscosity of the fluid, d is the diameter of the tube, and L is the distance over which the pressure drop is measured. The slope of the graph of the logarithm of ΔP versus the logarithm of Q (Fig. 2) should be unity in this region.

As the speed of flow is increased there is a transition region and then the turbulent flow regime is encountered. The corresponding expression to (1) for high flow speeds is Darcy's formula

$$\Delta P = \frac{32 f \rho L}{\pi^2 d^5} Q^2$$

where ρ is the density of the fluid and f is a friction factor, determined by the surface roughness of the pipe. In this region the slope of the graph in Fig. 2 should be 2. (In fact, Darcy's formula is exact only for rough pipes. A figure of approximately 1.8 is expected for a smooth pipe, as in this experiment.)

In addition to this analysis the mean flow speed $u = Q/\pi d^2/4$ can be expressed in terms of the dimensionless Reynolds number, given by

$$\mathrm{Re} = \frac{\rho d}{\mu} u$$

The onset of turbulence (point B, Fig. 2) occurs at a critical flow speed corresponding to the critical Reynolds number, which should be approximately 2000 for any diameter of pipe and any fluid.

A typical set of results obtained with this apparatus is shown in Fig. 3 where the volume flow rate has been varied between 10^{-5}

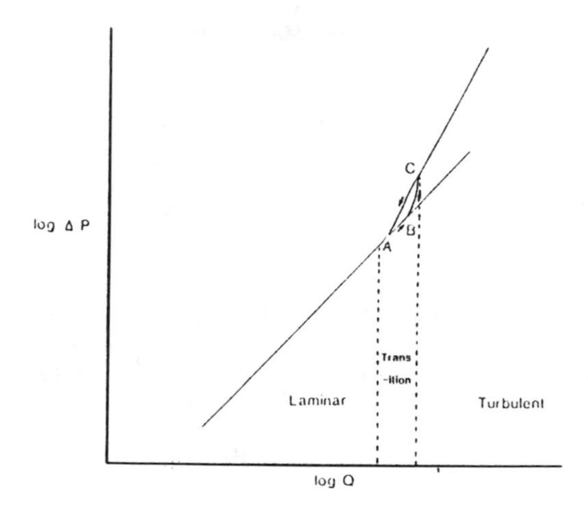

FIG.2 Theoretical plot of the logarithm of ΔP versus the logarithm of Q showing the two flow regions (laminar and turbulent) and a transition region

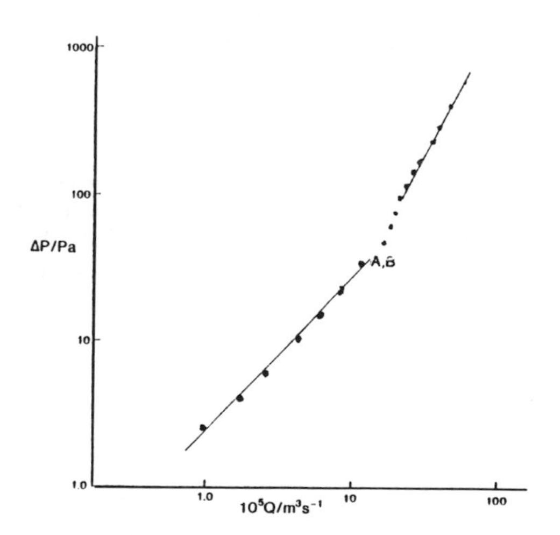

FIG.3 Experimental data showing a plot of ΔP versus Q on logarithmic scales. Students find it easier to plot the logarithm of ΔP and Q on a linear scale as the slopes of the two sections can then be more readily calculated

and $10^{-3}\,\mathrm{m^3 s^{-1}}$ (corresponding to flow speeds of 0.35 to $35\,\mathrm{m\,s^{-1}}$). The slopes of the two straight-line sections of the graph are 0.96 and 1.85, respectively, and the critical Reynolds number (measured at point A on Fig. 3) is 2060. In addition the scaling factors between ΔP and Q (laminar) or Q^2 (turbulent) are as expected. All these values are reproducible and accurate to 5%.

It has been found impossible to reproduce all features of Fig. 2 with this apparatus. In particular the different paths, A to B and C to A cannot be obtained. This is primarily due to turbulence introduced into the system by the flow-meters and other discontinuities, which ensures an early onset of turbulence as the flow rate is increased. Modifications of the apparatus are being considered, including sucking air through the tube so that the flow-meters and any regulator valves are downstream from the length of pipe over which the measurements are made, thus reducing the introduction of turbulence into the system. Apart from this, the slopes of the experimental data graph (Fig. 3) in the two flow regimes and the critical Reynolds number are well within accepted values. Students find this experiment interesting as it demonstrates some of the basic physics of fluid flow and produces reliable and reproducible results.

Streamline flow around an object

D.M. LIVESLEY, J.J. HARTLEY, AND J.J. LOWE
Department of Physics, Exeter University, Exeter, EX4 4QL, UK

Abstract: We describe a simple method of observing the pattern of streamlines around a two-dimensional object. Semi-quantitative measurements of the lift force can be made (indirectly) from these patterns, and examples of the results obtained are shown.

Introduction

Diagrams in elementary textbooks showing streamline flow around objects such as aerofoil sections are rarely accurate and often misleading. Such diagrams serve to reinforce the widespread, but false, idea that lift is due to the fluid flowing faster over one side of the aerofoil 'because it has further to go', supposedly giving rise to lift through a kind of Venturi effect. This explanation ignores the esssential fact that in order for there to be lift, there must be circulation around the aerofoil. The direction of lift is dictated by the sense of circulation. This in turn can be chosen either by giving the section a suitably asymmetric shape with respect to the fluid flow, as in the cases of wings and sails, or by causing a symmetric shape to rotate, as in the case of spinning balls. The ability of planes to fly either way up, and of sails of negligible thickness to provide motive force, can only be understood if this point is appreciated.

We describe simple apparatus that provides laminar flow in a thin sheet of water. The streamlines are easily made visible using dye streaks obtained by sprinkling a few crystals of potassium permanganate into the water. Inserting suitable objects in the flow to observe the effect on the streamlines is then a simple matter. Semi-quantitative measurements can be made if the streamline pattern can be photographed.

The apparatus used is shown in Fig. 1. A large $(12'' \times 16'')$ photographic developing tray is used to contain the water, and is tipped at a small angle $(1^\circ - 2^\circ)$ to ensure that the water overflows along one edge only. A carefully-fitted aluminum sheet provides a table on which objects can be mounted, and also directs the water flow into a thin sheet close to the surface. The table should, of course, be horizontal when the tray is tipped, so as to be parallel to the water surface. It should be positioned so as to provide water depths

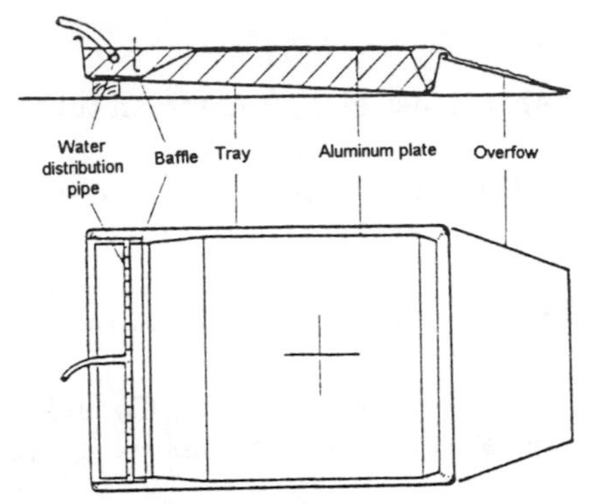

FIG.1 Top and side views of the flow tank. Objects are positioned close to the centre of the table (marked by a cross) to minimize edge effects

of a few millimetres. It is important to provide a well-distributed flow of water, and so the water supply is connected to a distribution tube with holes all along its length. This is situated behind a baffle plate, placed parallel to the top edge of the aluminum table, in order to minimize the amount of turbulence reaching the table. Water velocities of a few centimetres per second are suitable and are conveniently achieved with this arrangement.

Measurement of lift force

The lift force experienced by an aerofoil section a few centimetres long from the slowly-moving water is too small to measure easily (it is of order 10^{-4} N). However, it can be estimated from measurements of the variation in streamline spacing along either side of the section. Such measurements are adequate for comparing the effectiveness of various angles of attack or shapes of sections. The required theory is easily derived from Bernoulli's equation and the continuity equation, if it is assumed that the water is incompressible and that its depth is constant.

Bernoulli's equation states that

$$P + \frac{1}{2}\rho v^2 \tag{1}$$

is constant along a streamline. The local pressure is P, and ρ and v are the local fluid density and speed respectively. If a narrow streamtube of cross-sectional area A is considered, continuity requires that

$$vA\rho = \text{const along the streamtube} \tag{2}$$

If it is now assumed that the water is incompressible and has constant depth, it follows that

$$vx = \text{const along the streamtube} \tag{3}$$

where x is the width of the streamtube. Considering sections at two points along the streamtube and combining equations 1 and 3 gives

$$P_1 + \frac{1}{2}\rho v_1^2 = P_2 + \frac{1}{2}\rho(v_1 x_1/x_2)^2 \tag{4}$$

Apply this now to streamtubes on either side of the object, and let the streamtube widths be a_i and b_i at position i respectively. Also let position 1 be sufficiently far upstream that the flow is undisturbed: then P_1 and v_1 are the same for both streamtubes. The difference in pressure across the aerofoil is then

$$\Delta P = k\left(\left(\frac{a_1}{a_2}\right)^2 - \left(\frac{b_1}{b_2}\right)^2\right) \tag{5}$$

where

$$k = \frac{1}{2}\rho v_1^2 \tag{6}$$

Fig. 2 shows how this result can be applied to a streamline pattern, and Fig. 3 shows the resulting graphs of ΔP against position for a sail section at various angles of attack. By finding the area under these graphs, the lift forces generated can be compared, and the results are shown in Fig. 4.

It may be noted that the constant k is of order unity, and so ΔP is typically less than 1 Pa. The resulting changes in the water depth are therefore of order 0.1 mm, and the additional errors thus introduced should be negligible.

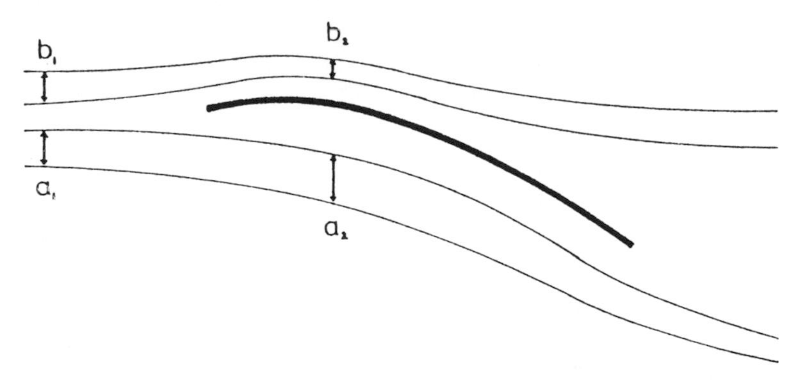

FIG.2 Two streamtubes are chosen, one on either side of the aerofoil, and their widths are measured at various positions along their lengths. Equation (5) can be used to calculate pressure differences between the two streamtubes from these measurements

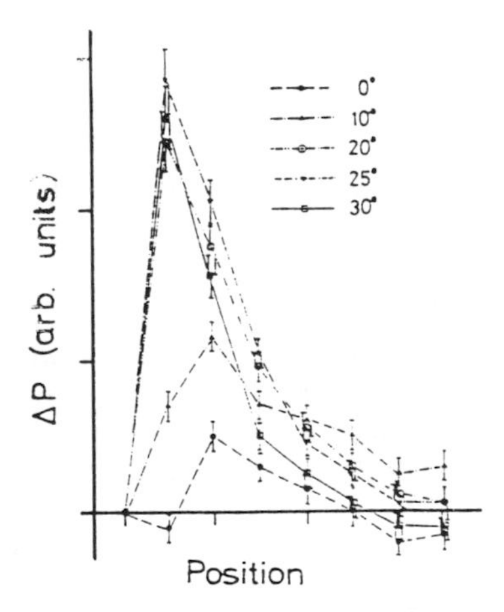

FIG.3 Graphs of ΔP against position along a sail section for different angles of attack

The Magnus effect

Circulation is established around an aerofoil in a fluid stream as a result of its asymmetric shape. Circulation can be established around a cylinder by rotating it in the stream, and this leads to a sideways force – the Magnus force – just as it leads to lift on an

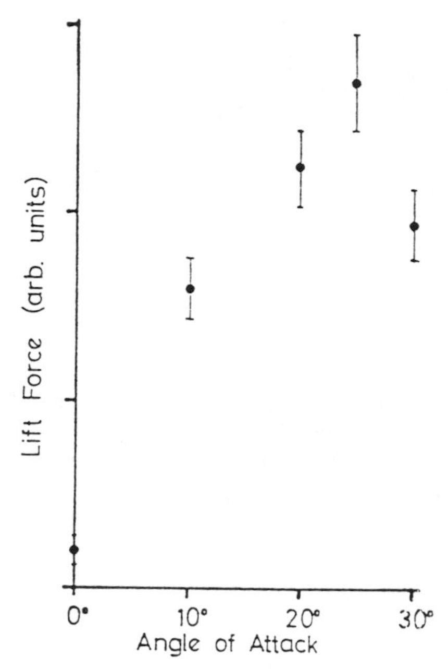

FIG.4 Lift force against angle of attack calculated from the results plotted in Fig. 3

aerofoil. This is what causes a spinning ball to swerve in flight. The corresponding distortion of the streamlines can be demonstrated using the apparatus described, and measurements of the Magnus force could be made using the above technique. Direct comparison of the results with theory should be possible.

The most satisfactory arrangement for this demonstration is to use a thick disk, which is positioned with its lower face as close as possible to the aluminum table, and which extends upwards through the surface of the water. For a 5-cm diameter disk, rotation at about 1Hz is fast enough: a rather slow speed is necessary to produce laminar flow with minimal shedding of vortices. Such speeds can be provided by a small electric motor driving through a gear box. The inevitable gap underneath the disk will upset the flow pattern slightly, but observations suggest that if this gap is made small, the effect will only be significant very close to the edge of the disk.

Tetragonal-cubic phase transition in barium titanate

D.M. NICHOLAS AND J.R. HORDER
School of Materials Science and Physics, Thames Polytechnic, London, SE18 6PF, UK

Abstract: Four different methods (X-ray diffraction, differential thermal analysis, optical microscopy and dielectric measurements) are described for the determination of the tetragonal to cubic transition in barium titanate ($BaTiO_3$). The series of experiments serves to demonstrate the relationship between crystal structure and physical properties, and is useful in integrating subject areas such as crystal chemistry, electrical, thermal and optical properties that are often studied separately.

Introduction

The series of experiments described here forms part of the second year undergraduate programme for a four-year sandwich Honors Degree in materials science. The production of the starting material, the examination of the phase change by a variety of techniques, and an understanding of the symmetry change consequent upon the phase change call upon concepts taught in a range of seemingly diverse courses such as crystal chemistry, X-ray diffraction, electrical properties, and crystal growth. Whilst the integration of these concepts is of value in illustrating the philosophy of materials science as an academic discipline, we would not wish to suggest that the experiments must be performed as a group, and doubtless the individual experiments could form the basis of more detailed studies in their own right.

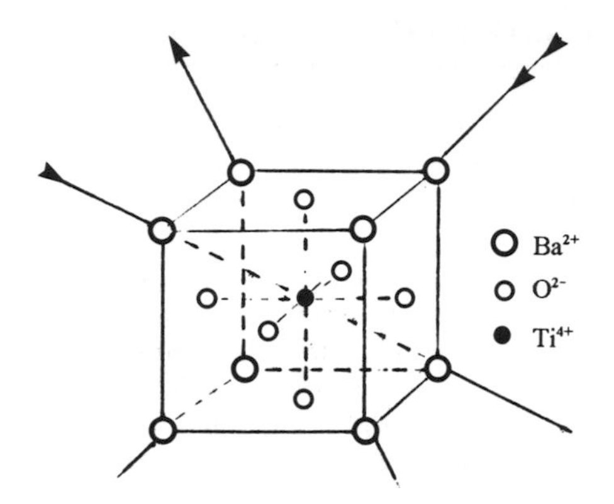

FIG.1 The cubic structure of $BaTiO_3$ above the ferroelectric Curie point

The crystal structure and the phase transition

Barium titanate ($BaTiO_3$) undergoes a number of phase transitions as its temperature increases. The one that forms the basis of this work is the tetragonal to cubic transition at about 120°C. In both forms, the structure can be considered as a cubic close packing of Ba^{2+} and O^{2-} ions with the Ti^{4+} at the body center (Fig. 1).

Above the transition temperature the structure is truly cubic, and the titanium ion *must* occupy the body center, at the intersection of the four body diagonals that are three-fold symmetry axes. Below the transition temperature, the cell suffers a slight tetragonal distortion ($c/a \approx 1.01$). Although the distortion is slight, the symmetry consequences are considerable. The three fold body diagonal symmetry vanishes, leaving a four fold axis parallel to c through the body center. Now the only requirement is that the Ti^{4+} lies on this axis. (Fig. 2).

If the Ti^{4+} is forced to leave the body center, this will cause polarization, and the origin of the piezoelectric property is thus apparent.

It is important to emphasise that it is the change of symmetry and not the size of the tetragonal distortion that is responsible for the piezoelectric property. Indeed, a calculation of the size of the unit cell based on the ionic radii of Ba^{2+}, O^{2-} and Ti^{4+} (1.43 Å, 1.33 Å and 0.64 Å, respectively) shows that the lattice parameter(s) in either form cannot be far from 4 Å.

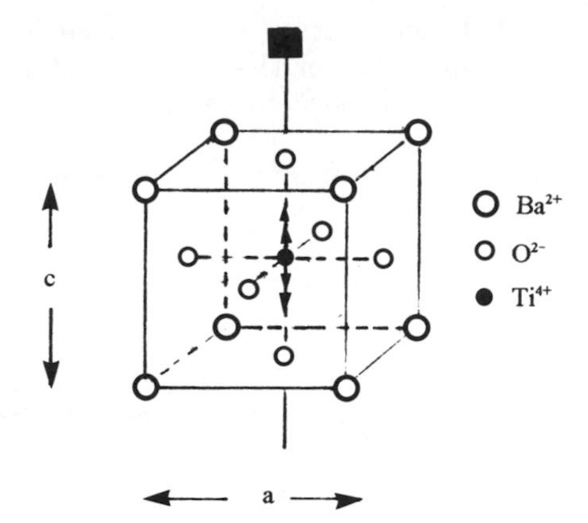

FIG.2

The materials

The material is to be subjected to examination by X-ray diffraction (XRD), differential thermal analysis (DTA), optical microscopy and dielectric measurement. Whilst commercially available, $BaTiO_3$ is satisfactory for XRD and DTA measurements, single crystal specimens are needed for the microscopic and dielectric measurements.

A number of methods of growing single crystals have been described. We have found the methods described by Remeika[1] and Eustache[2], involving the use of a KF flux, to be quite straightforward. Remeika's method is particularly useful in that it produces 'butterfly twins', fragments of which are very suitable for dielectric measurements.

Optical microscopy

Cubic crystals are optically isotropic and will be dark when viewed between crossed polars, regardless of their orientation to the polars, whilst tetragonal crystals are optically uniaxial and will in general, only exhibit extinction at four points in a complete rotation under crossed polars.

A small crystalline fragment of material is mounted on the hot stage of a polarizing microscope, and its position adjusted so that it gives a bright interference colors under white light illumination at room temperature. As the temperature rises, the colors persist until, at the transition temperature, the colors vanish abruptly.

The indicated temperature of the stage is clearly only very roughly the temperature of the crystal, but it is quite straightforward to calibrate the system using standard melting point materials.

Very small crystalline fragments exhibit much sharper changes than larger crystals.

DTA measurements

The DTA measurements are quite straightforward. The endotherms are not very large and it is as well to use the instrument on its most sensitive range. If the sample is cooled to −100°C it is possible to detect two other phase changes, one about −70°C and the other at about +5°C as well as the major change at 120°C. The two lower temperature transitions exhibit some slight hysteresis, depending upon whether the temperature is rising or falling.

A typical trace is exhibited in Fig. 3.

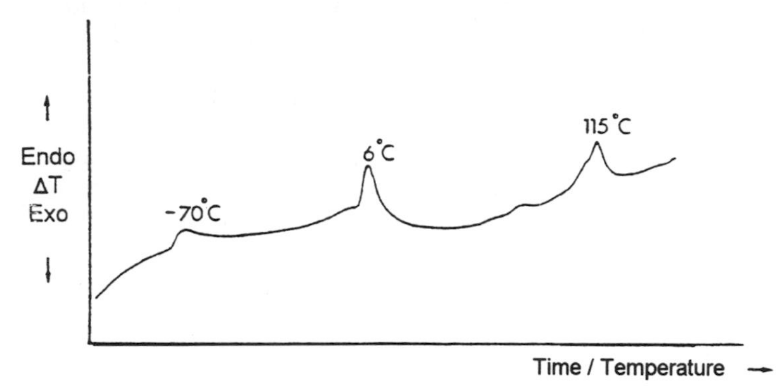

FIG.3 DTA of $BaTiO_3$

Dielectric measurements

The dielectric constant of the material exhibits an anomaly at each phase transition. A small fragment of one of the 'butterfly twin' crystals (approximately 3 mm × 2 mm × 0.5 mm) has been coated with silver dag on its two larger surfaces and leads fixed to them. It has been mounted in a small test tube with a thermocouple junction. The leads are taken out to a capacitance meter whilst, for convenience, the thermocouple temperature is read directly. The test tube is mounted in an oil bath and the capacitance, which is, of course, directly proportional to the dielectric constant, monitored as a function of temperature.

The anomaly in capacitance is clearly visible in Fig. 4.

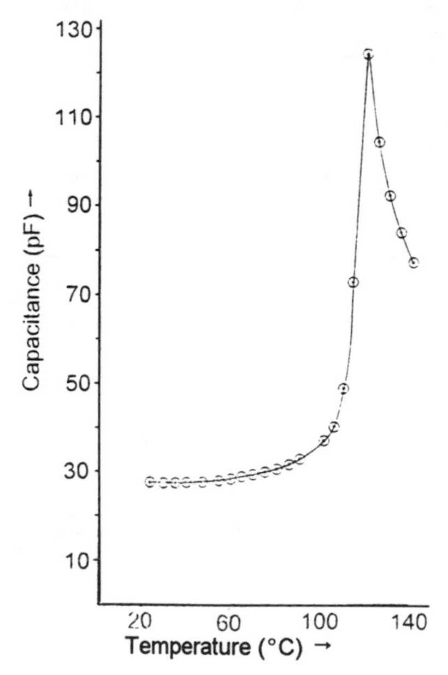

FIG.4 Dielectric measurements. Typical results for the capacitance of a crystal device as a function of temperature

X-ray diffraction data

Powder X-ray diffraction data has been obtained from a conventional X-ray diffractometer, fitted with a hot stage. The hot stage is a standard aluminium sample holder, to the back of which has been fitted a 50-W heating element taken from a soldering iron. This is driven by a Variac and can reach temperatures of about 250°C, well above those required here.

Displayed in Fig. 5 are comparisons of the traces from the material above and below its transition temperature. The high tem-

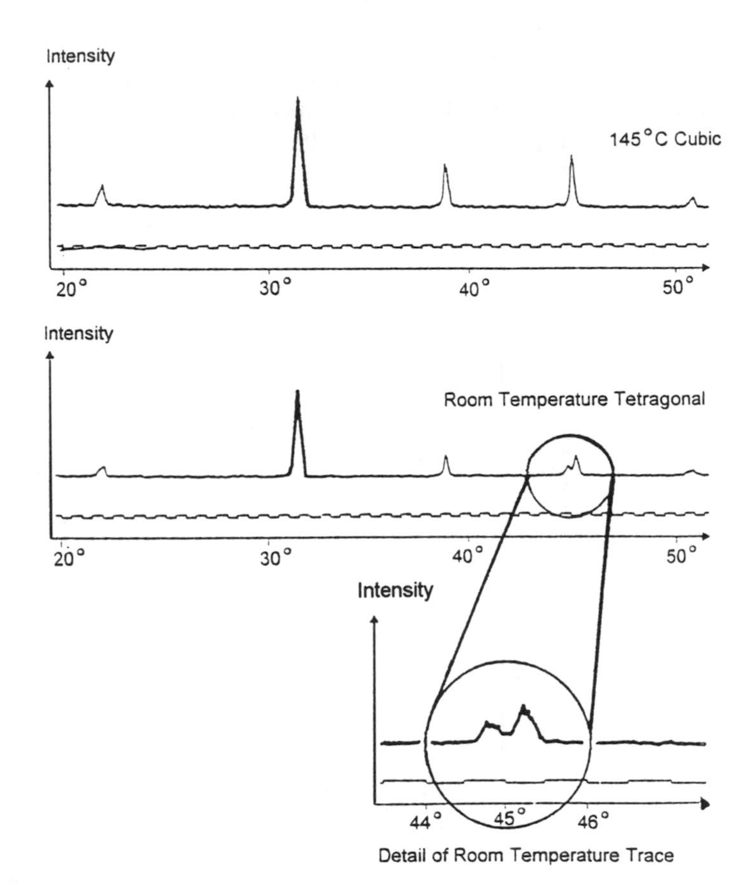

FIG.5 X-ray diffraction traces of $BaTiO_3$

perature (cubic) form gives a pattern characteristic of a primitive cubic cell and lines may be indexed on the basis that they represent consecutive values of $h^2 + k^2 + l^2$, starting from 1. The lattice parameter may be estimated by a standard extrapolation technique (Fig. 6). The significant difference exhibited by the low temperature tetragonal form is the tendency for the lines to split into doublets (or even triplets at high Bragg angles). This arises because the spacing d is given by

$$d_{hkl} = \left[\frac{h^2 + k^2}{a^2} + \frac{l^2}{c^2} \right]^{-1/2}$$

rather than

$$d_{hkl} = \left[\frac{h^2 + k^2 + l^2}{a^2}\right]^{-1/2}$$

as it was in the cubic case. Whereas the order of a set of indices *hkl* is immaterial to the spacing *d* in the cubic case, it *does* make a difference in the tetragonal case. For example, the fourth line in the tetragonal phase is the doublet d_{200}, d_{002} with spacings of a/2 and c/2, respectively, and these can be used to calculate *a* and *c*.

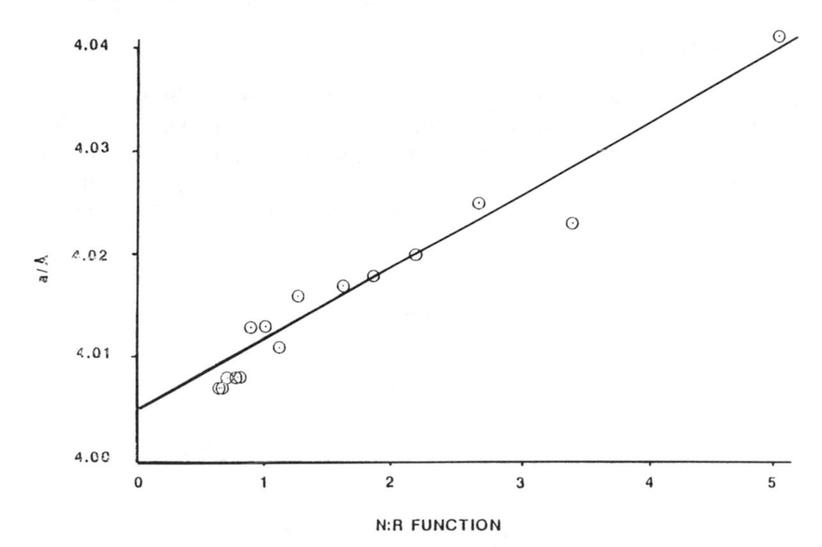

FIG.6 Lattice parameter versus Nelson–Riley function. $BaTiO_3$ > 120°C, $a = 4.005$ Å

It is, of course, good practice to obtain more accurate lattice parameters from the higher Bragg angle reflections. Considerable care is necessary if this is done with the tetragonal case as it is often quite difficult to distinguish the splitting of lines due to c/a from that due to $\alpha_1\alpha_2$ splitting.

References

1 J.P. Remeika, *J. Am. Chem. Soc.* **76**, 940 (1954)
2 H. Eustache, *Comp. Rend.* **244**, 1029 (1957)

Rainbow experiments

C. ISENBERG
Department of Physics, University of Kent at Canterbury, Canterbury, UK

Abstract: The experiments described here were originally set at the International Physics Olympiad that was held at Harrow School. A complete discussion of the theory of the rainbow has been given by Walker[1].

Introduction

The rainbow that one normally observes in the sky – the first-order rainbow – is due to light entering spherical raindrops and being internally reflected *once* before emerging from them as indicated in Fig. 1. The angle of incidence i at which greatest intensity of the light is observed to emerge is that associated with the angle of minimum deviation of the ray.

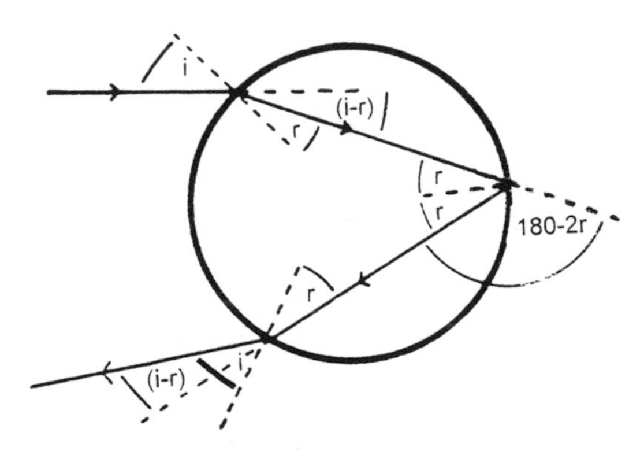

FIG.1 Origin of the rainbow

If, generally, i is the angle of the incidence and r the angle of refraction for a light ray entering a drop, the angle of deviation ϕ_1 is given by

$$\phi_1 = 180^\circ + 2i - 4r \tag{1}$$

Minimum deviation occurs when

$$\frac{d\phi_1}{di} = 0 \tag{2}$$

or, using Snell's law for a drop of refractive index n, when

$$\cos i = \sqrt{\frac{1}{3}(n^2 - 1)} \tag{3}$$

The second-order rainbow can frequently be observed in the sky together with the first-order rainbow. This occurs after two internal reflections of a ray of light in the drop. The angle of deviation of this ray is

$$\phi_2 = 2(180^\circ) + 2i - 6r \tag{4}$$

The minimum deviation condition, for which the greatest intensity is observed, is

$$\frac{d\phi_2}{di} = 0,$$

Again, using Snell's law, we have

$$\cos i = \sqrt{\frac{1}{8}(n^2 - 1)} \tag{5}$$

The intensity of higher-order rainbows, due to $3, 4, \ldots$ internal reflections in the raindrop, decreases with order. In daylight, one cannot observe the third and fourth orders as one has to look in the direction of the Sun; the intense light from the Sun makes it impossible to observe these weak bows. However, the fifth-order rainbowoccurs, like the first and second-order rainbows, in the direction away from the Sun, but is too weak to be observed against the background of diffuse white light. Consequently, only the first two rainbows are commonly observed in daylight.

In the laboratory, pendant drops of water, or any transparent liquid, can be used toobserve the higher-order rainbows. The deviation ϕ_k produced by a ray of light internally reflected k times in a drop is given by

$$\phi_k = k(180^\circ) + 2i - 2r(k + 1) \tag{6}$$

The minimum deviation condition for the kth-order rainbow together with Snell's law gives the general result

$$\cos i = \sqrt{\frac{(n^2 - 1)}{k(k + 2)}} \tag{7}$$

This introductory information should provide the background analysis necessary to understand the Physics Olympiad question. It should be pointed out that, when observations are made on pendant drops, using apparatus provided, one does not measure ϕ_k, but an angle $0 \leq \theta_k \leq 180^\circ$. One is required to deduce ϕ_k.

The Olympiad examination question

Apparatus (Fig. 2)

1. Spectrometer with collimator and telescope
2. Three syringes; one for water, one for liquid A and one for liquid B.
3. A beaker of water plus two sample tubes containing liquids A and B
4. Three retort stands with clamps
5. 12-V shielded source of white light
6. Black card, plasticine, and black tape
7. Two plastic squares with holes to act as stops to be placed over the ends of the telescope, with the use of two elastic bands
8. Sheets of graph paper
9. Three dishes to collect water plus liquids A and B lost from syringes.

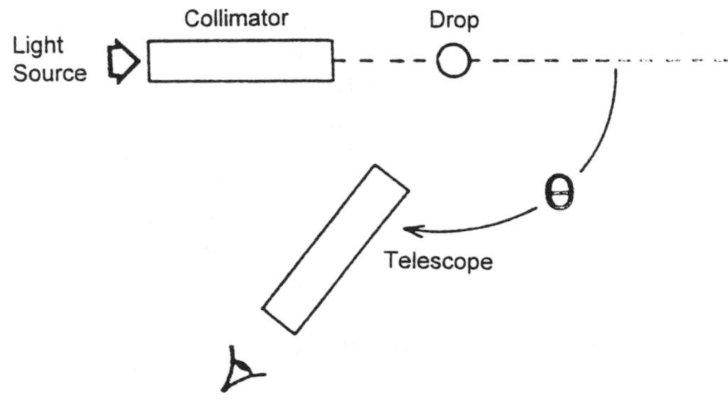

FIG.2 Plan of apparatus

Instructions and Information

1. Adjust collimator to produce parallel light. This may be performed by the following sequence of operations:
   (a) focus the telescope on a distant object, using adjusting knob on telescope, so that the cross hairs and object are both in focus
   (b) position the telescope so that it is opposite the collimator with slit illuminated so that the slit can be viewed through the telescope
   (c) adjust the position of slit, using the knob on the collimator, so that it is in focus on the cross hairs of the telescope eyepiece

(d) lock the spectrometer table, choosing an appropriate 'zero' on the vernier scale, so that subsequent angular measurements of the telescope's position can conveniently be made.

2. Remove the eyepiece from telescope and place black plastic stops symmetrically over both ends of the telescope, using the elastic bands, so that the angle of view is reduced.
3. Open up collimator slit.
4. Use the syringes to suspend, vertically, a pendant drop symmetrically above the centre of the spectrometer table so that it is fully illuminated by the light from the collimator and can be viewed by telescope.
5. The central horizontal region of the suspended drop will produce rainbows as a result of two refractions and k ($k = 1, 2, \ldots$) internal reflections of the light. The first-order rainbow corresponds to one internal reflection. The second-order rainbow corresponds to two internal reflections. The kth-order rainbow corresponds to k internal reflections. Each rainbow contains all the colours of the spectrum. These can be observed directly by eye and their angular positions can be accurately measured using the telescope. Each rainbow is due to white light rays incident on the drop at a well determined angle of incidence; which is different for each rainbow.
6. The first-order rainbow can be recognised as it has the greatest intensity and appears on the righthand side of the drop. The second-order rainbow appears with the greatest intensity on the lefthand side of the drop. These two rainbows are within an angular separation of 20° of each other for water droplets. The weak intensity fifth-order rainbow can be observed on the righthand side of the drop located somewhere between the other two, 'blue', extreme ends of the first and second-order rainbows.
7. Light reflected directly from the external surface of the drop and that refracted twice but not internally reflected, will produce bright white glare spots that will hinder observations.
8. The refractive indices of the liquids are:

$$\begin{aligned} \text{Water} \quad n_W &= 1.333 \\ \text{Liquid A} \quad n_A &= 1.467 \\ \text{Liquid B} \quad n_B &= 1.534 \end{aligned}$$

A summary sheet is expected to be completed in addition to the experimental report.

Measurements

1. Observe, by eye, the first- and second-order water rainbows. Measure the angle θ through which the telescope has to be rotated, from the initial direction for observing the parallel light from the collimator, to observe, using a pendant water droplet, the red light at the extreme end of the visible spectrum from:
 (a) the first-order rainbow on the right of the drop ($k = 1$)
 (b) the second-order rainbow on the left of the drop ($k = 2$)
 (c) the weak fifth-order rainbow ($k = 5$), between the first and second-order rainbows.

 One of these angles may not be capable of measurement by the rotation of the telescope due to the mechanical constraints limiting the range of θ. If this is found to be the case, use a straight edge in place of the telescope to measure θ.

 (Place the appropriate dish on the spectrometer table to catch any falling droplets).

 Deduce the ***angle of deviation***, ϕ, that is the angle the incident light is rotated by the two refractions and k reflections at the drop's internal surface, for (a), (b), and (c). Plot a graph of ϕ against k.

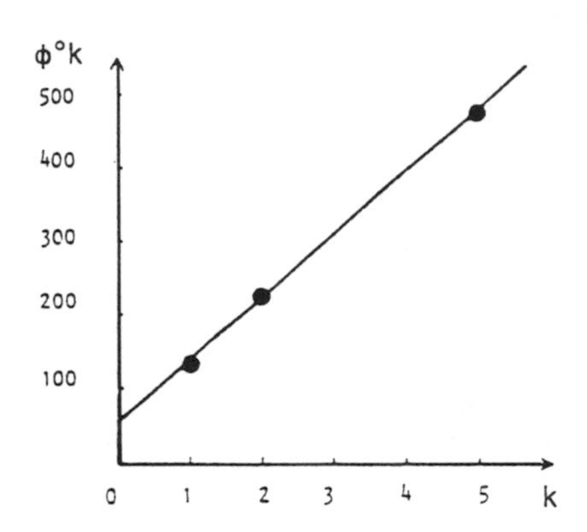

FIG.3

2. Determine ϕ for the second-order rainbows produced by liquids A and B, using the red visible light at the extreme end of the visible spectrum. (Place respective dishes on table below to catch any falling liquid as the quantities of liquid are limited).

 Using graph paper plot $\cos\phi/6$ against $1/n$, n being the refractive index, for all three liquids and insert the additional point for

$n = 1$. Obtain the *best* straight line through these points; measure its gradient and the value of ϕ for which $n = 2$.

Figures 3 and 4 show some typical results.

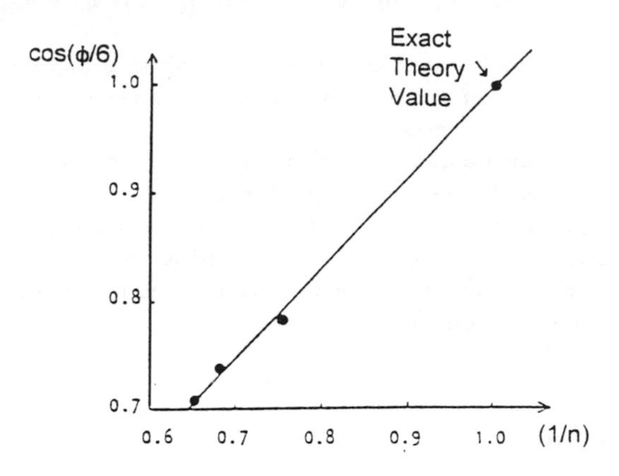

FIG.4

References

1 J.D. Walker, *Am. J. Phys.* **44**, 5, 421 (1976)

Computation of lens aberrations

J. SWAGE
*Software Development Unit, Department of Pure and Applied Physics, University of Salford, Salford, M5 4WT, UK**

Abstract: A great deal of knowledge of the physical and biological world comes from our use of microscopes, telescopes, cameras, and other optical devices that use light waves to form images of greater brightness or detail than we can obtain from our eyes alone. A basic part of the design of such optical systems is the tracing of rays through them and the determination of their deviations from perfect imagery. These deviations are called aberrations, the most important of which is the spherical aberration. The design of lenses involves many tedious calculations and can easily be facilitated by the production of a computer program.

Lens aberrations

The Software Development Unit in the Department of Pure and Applied Physics at the University of Salford has been set up on the basis of a grant awarded by the Computers in Teaching Initiative. It has developed high quality computer simulations for teaching undergraduates physics and computational physics. The software is written in strict standard FORTRAN 77 and uses the device independent Graphical Kernel System (GKS) for the graphical input/output. Two of the programs have been collected together to form a software package on the computer aided analysis of lens aberration[1] (see Refs. 2 and 3 for background reading).

Program 1[1] calculates the trajectories of the rays passing through a two-dimensional lens system by tracing the ray from one surface to the next, and calculating the refraction at each surface using Snell's law. Figure 1 shows an example graphical output from the program.

Two types of ray tracing are employed. Finite ray tracing involves an exact calculation as described above. Paraxial ray tracing is an approximate method in which it is assumed that the rays are traveling close to the optical axis, and nearly parallel to it, so that certain small angle approximations may be used.

---

* Now at N.N.C. Ltd., Knutsford, UK

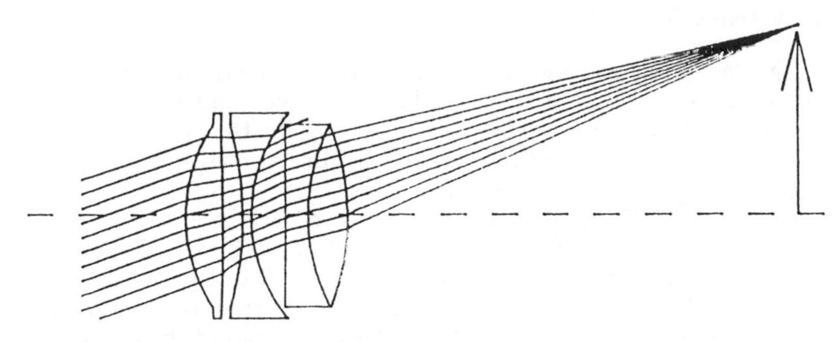

Fig.1 A plot of a Tessar photographic lens and a ray trace through it. The object is at infinity and the initial angle of incidence is 20°. The arrow indicates the position of the Gaussian image

Since lens properties such as focal length vary depending on the position of the incident ray relative to the optical axis, these properties are defined according to a paraxial ray trace, and the aberrations are defined according to the deviations from these values of the actual lens properties.

The most important aberration, which is present even for axial objects, is the spherical aberration that takes the general form of producing different focusing positions for different incident heights. The distance along the optical axis between the focus for the marginal ray and the focus calculated from the paraxial ray trace is the longitudinal spherical aberration, and the distance betweeen the two at the Gaussian image plane is the transverse spherical aberration. The program calculates and displays plots of the transverse spherical aberration as a function of incident height.

After an initial design has been set up, and its aberration calculated, the design can be altered so as to minimise the aberration. The simplest way to do this is to leave the glass types and separations unchanged, but to alter the curvatures of the surfaces in a systematic way, since the aberration varies quite strongly with this parameter. This can be done by a technique known as bending, whereby the curvatures of all the surfaces are modified so that the change in curvature at each surface is inversely proportional to the height. In this way the focal length of the lens remains constant. The user can investigate the variation of aberration with respect to the bending and can study the modified lens for a specified value of the bending parameter with the software that has been developed.

Ray tracing

Program 2[1] extends the ray trace to three dimensions. It allows the user to manipulate the lens data provided, and to look at a plot of the lens system. The user can then plot the spot diagram, the modulation transfer function and the line spread function. All these are widely used in designing lenses.

It is often difficult to decide whether the performance of a specific lens is sufficient for a given application. The usual method is to perform a three-dimensional ray trace of a large number of rays from a point source in a uniformly distributed array over the entrance pupil of the lens, and then plot a diagram of the points at which these rays pierce the image plane (the spot diagram). The shape of the diagram and the density of the points will depend on the position of the chosen image plane and on the aberrations of the system. In the limit of an infinite number of rays the point density tends to what is known as the point spread function. The spot diagram is obviously a finite approximation to the point spread function to which it will approach more nearly as the number of rays traced increases.

The line spread function is obtained by holding one of the two-dimensional image coordinates constant and integrating over the other. Thus, whilst the point spread function is the image of a point in the object plane, the line spread function is the image of a line. Since most lenses have rotational symmetry, the line spread function is sufficient to describe the characteristics of an image near the axis.

In considering the characteristics of a lens system, the amount of contrast that can be transferred to the image is important. It is usual to define the contrast of a wave function as

$$\mathrm{Modn} = \frac{I_{\mathrm{max}} - I_{\mathrm{min}}}{I_{\mathrm{max}} + I_{\mathrm{min}}}$$

The transfer factor, or modulation transfer function, which is a function of the spatial frequency, is given by

$$M(\nu) = \mathrm{Modn(Image)}\,\mathrm{Modn(Object)}$$

One way to interpret the significance of a spot diagram is to regard it as a point spread function and to Fourier transform it into a curve of the modulation transfer function plotted against spatial frequency. Such a graph contains information both as to the resolving power of the lens and the contrast in the image of coarse objects and, since diffraction effects can be taken into account, the result is the most comprehensive representation of lens performance that can be obtained.

References

1 *Computer Aided Analysis of Lens Aberrations*, a software package available on request from the Software Development Unit (CTI), Department of Pure and Applied Physics, University of Salford, Salford, M5 4WT, UK
2 Rudolf Kingslake,*Lens Design Fundamentals*, Academic Press, 1978
3 A.D. Boardman (ed.),*Physics Programs*, John Wiley, 1980

Fourier transform spectroscopy

J.D. Collins and T.J. Parker
Department of Physics, Royal Holloway and Bedford New College, Egham, Surrey, TW20 OEX, UK

Abstract: In this experiment a Michelson interferometer is used to illustrate the principles of Fourier transform spectroscopy to students in the third year laboratory. Interference fringes produced by simple light sources such as broad band tungsten lamp with color filters, a sodium lamp, and a He-Ne laser are detected with a silicon photodiode, amplified, and Fourier transformed with a Tuscan microcomputer. The instrument can be used to determine the spectra of these light sources and to investigate the relationship between the interferogram and the spectrum. More quantitative investigations that require more advanced work include the study of resolution, apodisation, and aliasing.

Introduction

The technique of Fourier transform spectroscopy (FTS) was originally confined to the far infrared region of the spectrum. However, during the past twenty-five years, as microcomputers have become cheaper, faster, and more powerful, it has gradually superseded more conventional methods at shorter wavelengths. The advantage of greater speed, together with the extremely high wavelength accuracy that can now be achieved with laser controlled sampling techniques, has resulted in the development of instruments to cover the whole spectral range up to the vacuum ultraviolet.

The principle of the Michelson interferometer is illustrated in Fig. 1. A beam of light from the source S is divided into reflected and transmitted partial beams at the front surface of a beam divider B, is coated with a semi-transparent metal film. The two partial beams are subsequently reflected by mirrors M1 and M2 and pass back to the beam divider where they are recombined at the front surface to form the output beam that is observed at the detector D. One of the mirrors (M2 in this case) can be moved in the direction normal to the incident light beam, and the aim of the experiment is to observe interference fringes by varying the optical path difference when M2 is scanned. Since the beam transmitted to M2 passes through B three times while that reflected to M1 passes through it only once, a compensator C, carefully matched

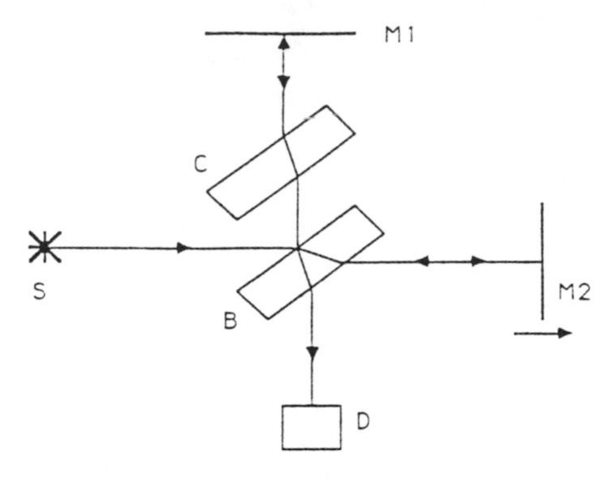

FIG.1 Optical arrangement of a Michelson interferometer: S – light source, B – beam divider, C – compensator, M1 – fixed mirror, M2 – moving mirror, D – detector

in thickness to the beam divided, is inserted in one arm to restore the balance in the optical path lenghts of the two partial beams.

Theory of Fourier transform spectroscopy

Before using the instrument the student requires a basic understanding of the theoretical basis of FTS, including, for instance, the Fourier transform integral, the instrument function and apodisation, spectral resolution, discrete sampling, and a qualitative treatment of aliasing. This information, together with descriptions of the advantages of FTS over grating and prism spectroscopy, can be obtained from the many books that are available on the subject[1,2] and will not be repeated here.

The Fourier transform spectrometer

A Michelson interferometer from Glen Creston[3] is used for this experiment, and the complete Fourier transform spectrometer is illustrated schematically in Fig. 2. Two lenses are added so that the instrument can be operated with collimated light beams, and the output beam is focussed on a photodiode. The time constant of the data acquisition system is short, so it is necessary for all light sources to be powered by d.c. supplies, otherwise intensity

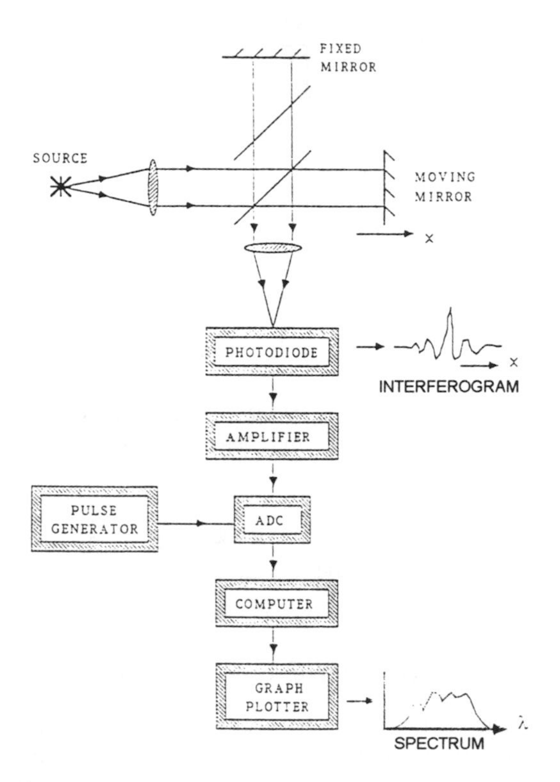

FIG.2 Schematic diagram of a Fourier transform spectrometer

fluctuations arising from the mains power supply would lead to spurious features in the computed spectra.

The output signal is amplified before it is passed to an analogue-to-digital convertor (ADC). The moving mirror is mounted on a slide connected to a micrometer screw, driven continuously by a d.c. motor, so that there is no direct means for determining the sampling interval for recording the interferogram. This is done by triggering the ADC with a pulse generator with a variable time base. Although this arrangement is acceptable for simple demonstrations, it should be noted that equal sampling intervals are only obtained if the screw moves at a constant speed. As shown later, deficiencies in the drive mechanism are clearly revealed if a laser is used as the source.

The computer is equipped with a fast Fourier transform (FFT) program and sub-routines for a number of standard Fourier transform operations that can be operated very easily by the student once the principles are understood.

Experiment

The interferometer is first aligned by eye, using a sodium lamp that produces an extended fringe system, and the position corresponding to zero optical path difference between the two beams is found, first approximately, by finding the central envelope of the Na fringes, and then more accurately by finding the white light fringes. The photodiode is then placed at the central fringe of the output beam and, after checking that the aperture is smaller than the diameter of the central fringe at all positions of optical path difference, the instrument is ready for use. The student then follows simple instructions on the monitor to record interferograms, Fourier transform them, and plot the resulting spectra.

Typical results

The FFT program requires that the following parameters are specified: the sampling interval, the number of points in the interferogram, whether or not apodisation is required, and the first and last wavenumbers in the output spectrum.

It is best to begin with a monochromatic source with a known wavelength, such as a He-Ne laser, as this can be used to calibrate the sampling interval for the rest of the measurements. Figure 3(a, b) shows a typical set of laser fringes and a laser output spectrum obtained after calibrating the wavenumber scale. Spurious structure arising from a combination of screw errors and non-uniform screw motion is clearly visible in the spectrum. The resolution of the measurement can be investigated by varying the length of the interferomgram and observing its effect on the linewidth of the output spectrum. Similarly, the instrument function can be investigated by computing spectra with and without apodisation.

The interferometer can also be used to measure broad-band spectra, as illustrated in Figs. 4 and 5. Figure 4(a) shows the strongly localised interferogram produced by a tungsten filament lamp. If the bandwidth is narrowed by adding a broad-band yellow (transmitting) filter, the interferogram becomes less localised, as shown in Fig. 4(b). The computed white-light spectrum, and the spectra

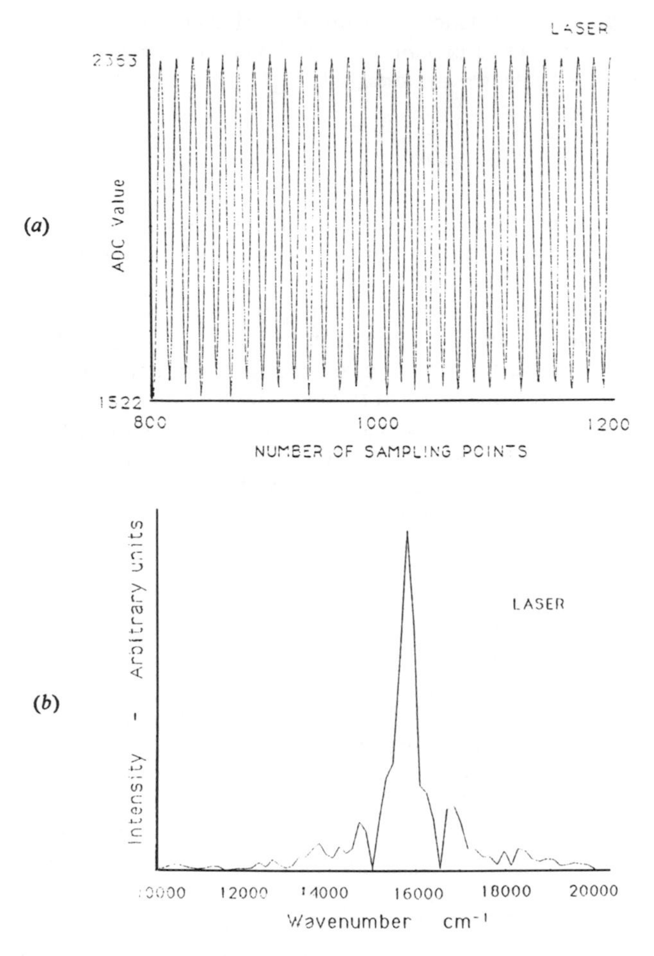

FIG.3 Portion of an interferogram recorded using a He-Ne laser for the source (a) and an example of a computed laser spectrum (b)

obtained by using the tungsten lamp with a series of broad-band color filters, are shown in Fig. 5(a, b). It can be seen that, compared with Fig. 3(b), no evidence of spurious structure is apparent in these spectra. This is a consequence of the localisation of the interferograms.

For more advanced work the instrument can be set up with a laser beam passing along the optical axis to monitor the path dif-

(*a*)

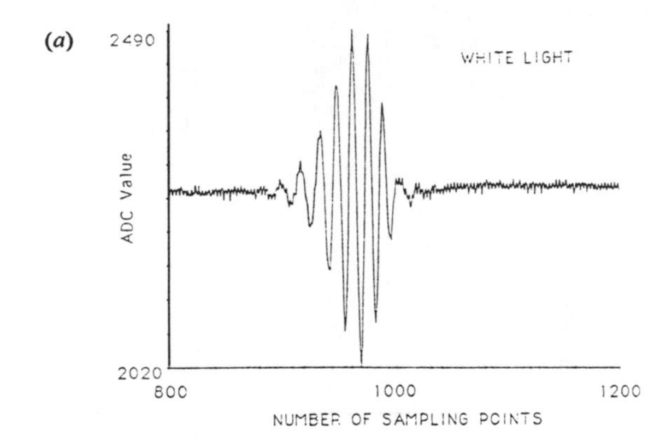

(*b*)

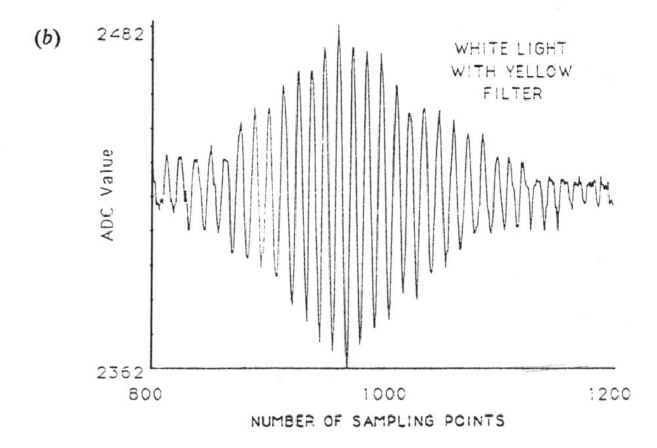

FIG.4 Localised interferogram recorded using a 12-V tungsten lamp for the source (a) and the slightly extended interferogram produced by limiting the bandwidth with a broad band yellow (transmitting) filter (b)

ference optically. This greatly reduces the effects of systematic errors in the drive mechanism and improves the sampling accuracy since the sampling interval is determined directly in terms of the laser fringe spacing, so that much better quality spectra can be obtained. The limitations of a small microcomputer can also be circumvented to some extent by recording interferograms with only a limited number of widely spaced points and using aliasing techniques to obtain spectra at much higher resolution.

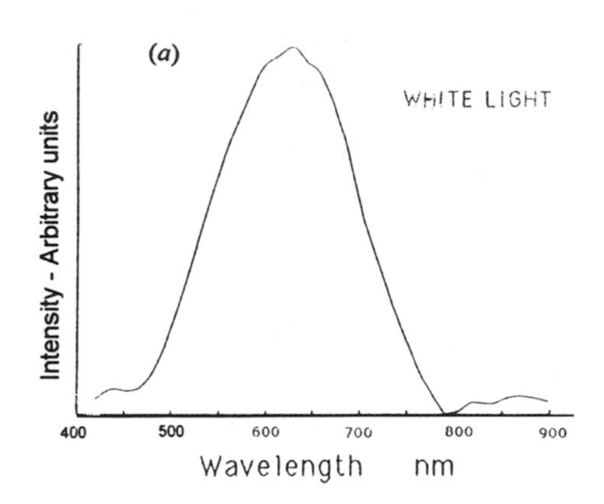

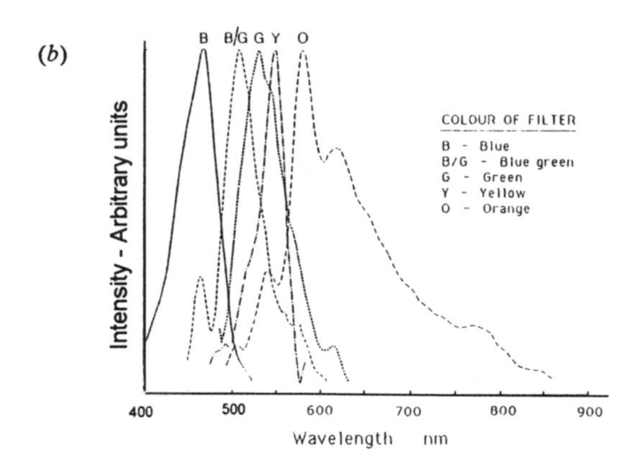

FIG.5 Calculated spectrum for the tungsten lamp (a) and a selection of computed spectra obtained with the tungsten lamp using different broad-band colour filters (b)

References

1 R.J. Bell, *Introductory Fourier Transform Spectroscopy*, Academic Press (1972)
2 D.C. Champeney, *Fourier transforms and their physical applications*, Academic Press (1973)
3 Glen Creston, 16 Carlisle Road, London, NW9 OHL, UK

Minimal holography

S. WOMACK, I. RUXTON, I. HARCUS,
AND G. FINLAYSON
*Department of Physics, Dollar Academy, Dollar,
Clackmannanshire, FK14 7DU, UK*

Abstract: A sixth-form project aimed at production of simple holograms of both the 'white light volume' and 'coherent light plane' types, all for a minimum capital expenditure. Totally student motivated, the project involved initial research, collection and construction of suitable equipment and the eventual practical production of holograms. The entire enterprise was undertaken in a short period and, following the current trend in scientific research, had practically no financial support whatsoever.

The first requirement being a vibration free environment, some research was conducted towards the simplest low cost approach, and a simple 'sand table' system decided upon. A large wooden box (approximately 1.6 m × 0.7 m × 0.2 m) forms the main part of the table, filled with around a hundredweight of sand, furnished by a nearby long-jump pit. The box is then supported on two partially inflated inner tubes, kindly donated by Kwik-Fit. Once assembled (see Fig. 1) on a study bench on a concrete floor, this basic set-up gave excellent isolation from vibration and an adjustable working surface suitable for our somewhat motley selection of equipment.

The most costly single item in the project is the laser, already owned by the physics department, and of the type becoming common in many schools. Costing around $150, the laser used for all our exposures is a simple multimode helium-neon with a power rating of < 1 mW. All optics were similarly obtained from around the school – microscope objectives and single lens elements of a reasonable, though not outstanding, quality, and a small front-silvered mirror. The only other important component, the plate holder, was built from wood to a simple design. All other items are trivial, e.g., clamps, stands, masses, etc., and are everyday equipment in any lab. Materials divide into two categories: the cheap, easily obtained, processing chemistry, and the expensive holographic plates themselves. The latter account for the only significant expenditure of the project. We used both 8E75 and 10E75

holographic plates themselves. The latter account for the only significant expenditure of the project. We used both 8E75 and 10E75 Agfa Holotest plates, the former being somewhat finer grained and yielding a gain in quality, but a significant loss in speed. Both retail at approximately $45 for a box of 20 plates, but a visit to Stirling University resulted in the very generous donation of enough plates to carry out our initial experiments. Holographers differ as to preferred chemistry, but we used Ilford PQ developer and HYPAM fixer. Processing times of 5 min development and 2 min fixation gave good results at room temperature. The resulting amplitude holgrams were not of sufficient clarity to give a good image, so bleaching was necessary to convert them to phase holograms. A simple 5% bromine solution, followed by thorough washing, achieves the necessary etching of the emulsion.

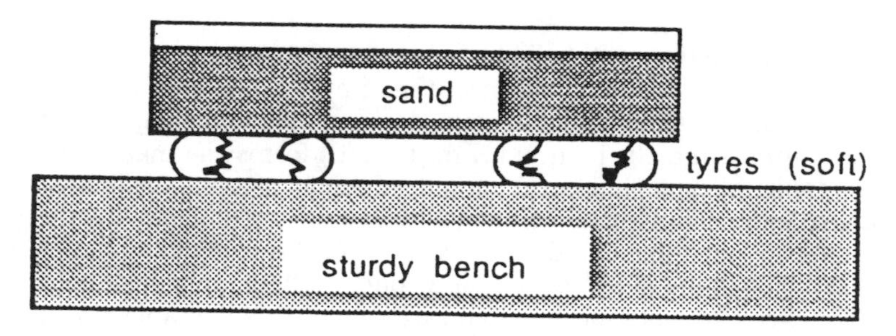

FIG.1 Simple isolation table

The first technique that we adopted was a simple method for producing white light reflection holograms, known as Denisyuk's method. This was chosen primarily for the simplicity of the setup involved (see Fig. 2). This is a great advantage, but there are disadvantages as well. Depth is limited because of the unavoidable difference in path lengths that is always twice the object-plate distance. Since the effective coherence length of the simple multimode laser involved is about 100 mm, the effective maximum depth is restricted to approximately 50 mm. Furthermore, we found that, in practice, the limit is usually nearer 30 mm. The other significant limitation is that it is impossible to control the relative intensities of the object and reference beams, since they are one and the same. It is the very ingenuity of Denisyuk's method that causes its limitations.

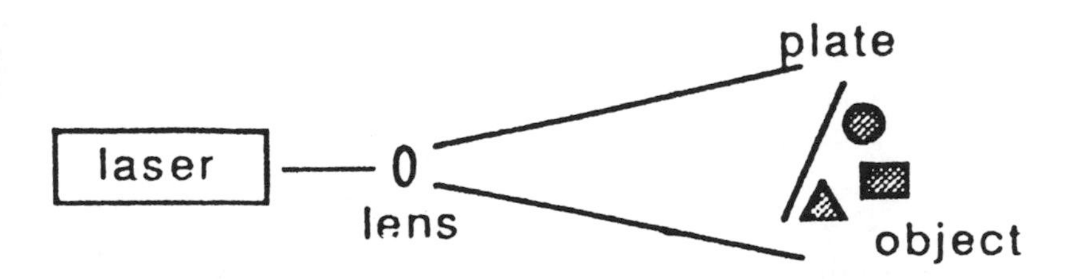

Fig.2 Denisyuk's method

Our experience showed that an important factor when using this technique is the reflectivity of the object. White or metal objects are best, but polished metal tended to show highlights rather than a complete image. Our best result was obtained using the mechanism of a clock, which was reflective without being highly polished. We realised, however, that there was no way in which we could improve upon our results with this technique, so the project took a new direction as we switched to coherent light transmission holograms.

Since no beam-splitter was available at this point, the simplest method of achieving coherent light transmission holograms was to use a single beam spread very widely (see Fig. 3). This method results in holograms of excellent depth and brightness but still afforded no control over the intensity ratio. Up to around 300 mm depth was achieved, and, after bleaching, bright, high-quality images were obtained. The quality of our results with this technique is significantly higher than that of those plates taken with Denisyuk's white light method, although the disadvantage lies in the somewhat inconvenient viewing requirements.

One particularly successful plate to emerge from this group of exposures was initially undertaken as a demonstration of the accuracy and 'reality' of a hologram; the object being a breadboard with various electronic components, and a strong magnifying lens in the foreground. The result shows a holographic lens, which coupled with the parallax effect, gives a most striking hologram for demonstration purposes, since the eye can view components magnified by looking through the lens and, with a small change in viewing angle, can see 'round' the lens so that the components are seen at their normal size.

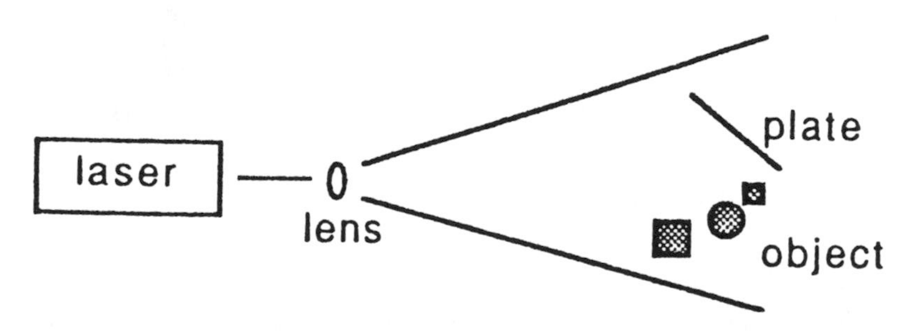

Fig.3 Single beam method for coherent light transmission holograms

In order to progress any further, it became clear that a beam-splitter was required. Much ingenuity was directed towards a low-cost method; plate glass at suitable angles, semi-silvered sunglasses and the like, but none of these produced beams of sufficient clarity and a reliable ratio between beams, so the realisation dawned that, for the first time in the whole project to date, a piece of commercial, professional, and probably expensive equipment would have to be obtained. However, enquiries resulted in the long-term loan of just such an item from Ferranti, Edinburgh. With this, we were able to implement the 'classic' method (see Fig. 4). As well as greater control over path-length and beam intensity ratio, this method gives much greater flexibility in object illumination. The result that we were able to achieve using this system were very rewarding, both in terms of depth and of brightness, which showed a gain over that of earlier single beam results, since path-length discrepancies can be almost totally eliminated. It was found to be a useful addition to the set-up if more mirrors are used, since the lighting of the object and the 'window' position can be easily changed to give whatever effect is desired, e.g., an image central on the plate.

As a step towards the application of holography, we undertook to produce a simple holographic interferogram, showing the dis-

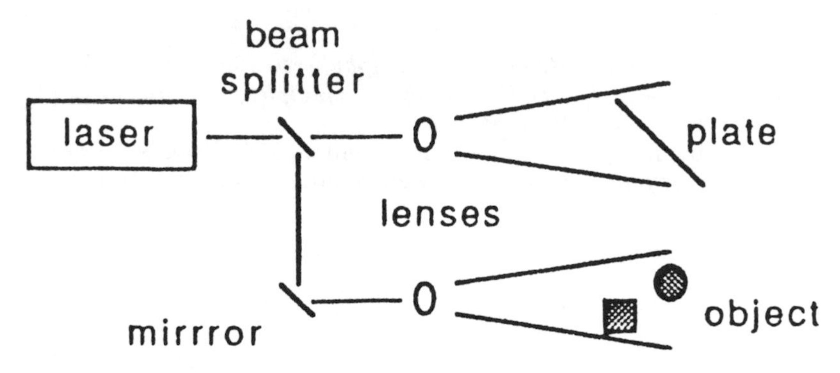

FIG.4 Twin beam method

tortion caused by a rubber band around a Coca-Cola can, by a double exposure method. Since this involved two very long exposures (8 minutes each) and the cutting of the rubber band inbetween, movement and vibration were obvious hazards. However, we achieved a satisfactory result after two trials; a tribute to our isolation table, if nothing else. Although very basic, holograms of this type are an obvious demonstration of the techniques now being used for such applications as non-destructive testing in industry, and damage analysis in art restoration.

As a result of this project, the school now possesses the basic equipment and knowledge to provide an ongoing sixth form holography project that has enormous scope for expansion and experiment with other techniques and processes. The authors also feel that this enterprise has proved the feasibility of such a project within an ordinary school, with little expense and no specialist equipment, despite the common assumption that holography is both very expensive and very difficult.

Further information may be obtained from Mr. P. Summut, Physics Department, Dollar Academy, Dollar, Clackmannanshire, FK14 7DU, UK.

References

1 M.Wenyon, *Understanding Holography*,Arco Publishing, 1978
2 P. Hariharan, *Optical Holography: Principles, Techniques, and Applications*, C.U.P., 1984
3 G. Saxby, *Holograms*, Focal Press, London, 1980 (This is the best text for anyone with little knowledge and even less equipment)

The Rijke effect

S.M. Kay
Royal Holloway and Bedford New College, University of London, Egham, Surrey, UK

Singing tubes – the Rijke effect

A metal cylinder with a wire gauze inside, fixed across the tube, is heated from one end. When the heat is removed the tube emits a sound not unlike that of a fog horn described as a 'howl', and this howl may last for several seconds.

The 'howling' effect was discovered by Rijke in 1859 and was explained by Lord Rayleigh in his 'Theory of Sound' in 1896. More recently, Ffowcs Williams and Dowling[4] have included an explanation of the effect in their book, but standard textbooks do not provide a detailed analysis of the phenomenon. The effect is therefore ideal for a student investigation, since there is no 'set' answer.

Experimental details

An iron or brass tube 5.5 cm in diameter with a 30 mesh iron gauze inside it works well, so well in fact that students working on this effect are often banished to the far side of the laboratory! (Lord Rayleigh used a cast-iron pipe 5 feet long and 4.75 inches in diameter hung from a beam in the roof of his laboratory!) An inner tube free to move up and down can be used to vary the total length of the tube from about 50 cm to 80 cm (see Fig. 1). The gauze is heated from the bottom of the tube with a Bunsen burner. The frequency of the howl can be measured by matching the sound to the output of an audio oscillator using beats. The temperature of the air just above the gauze and at the top of the tube can be monitored with thermocouples, and a noise level meter can be used to monitor the intensity of the sound emitted.

Things to investigate

Students can make some initial qualitative investigations to determine

(i) if the tube howls when it is horizontal
(ii) if it howls when the wire gauze is near the top of the tube
(iii) if it matters how hot the tube is.

Quantitative measurements can be made of

(i) the variation of the frequency with tube length,
(ii) the duration of the howl in relation to the temperature,
(iii) the effect of mesh size on the intensity of the howl.

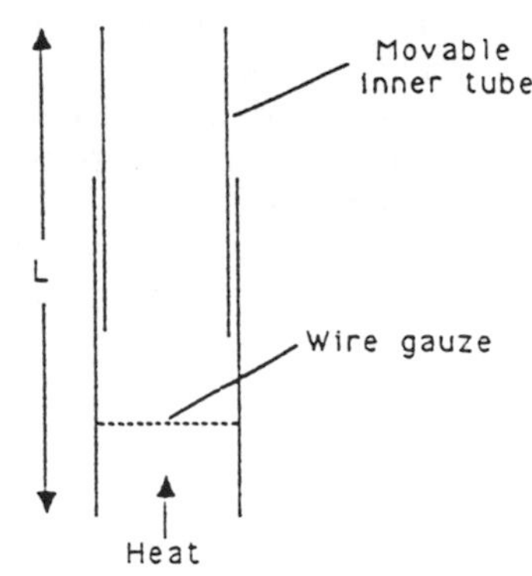

FIG.1 The Rijke tube

Some results

It is found that the tube must be vertical and open at both ends, and also that the wire gauze must be in the lower section of the tube for it to howl. With the heat supplied by a Bunsen burner there is a minimum length of tube required for howling, in this case about 53 cm.

The frequency f of the sound emitted corresponds to the fundamental frequency of an open pipe of length L; hence $f = V/2L$, where V is the velocity of sound at the air temperature T.

Typical data, plotted as $1/f$ versus L, are shown in Fig. 2. From the slope, a value of $403 \pm 44\,\mathrm{ms}^{-1}$ is obtained for the velocity. This is the velocity of sound in the hot air inside the tube, in this case corresponding to a temperature of 130°C (calculated taking

330 ms^{-1} for the velocity at 0°C and that assuming V is proportional to $\sqrt{T}$). In actual fact the frequency of the emitted note can be heard to decrease during the several seconds' duration of the howl as the air cools down.

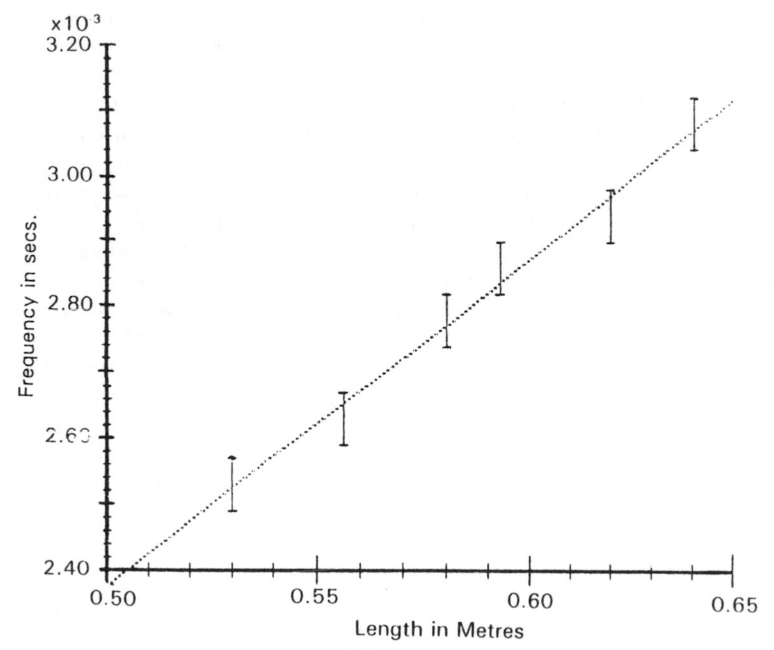

FIG.2 Graph of 1/frequency versus length of tube

By following the temperature changes in the tube, it is found that on removal of the heat the temperature at the top of the tube falls fairly quickly, but just above the gauze the temperature continues to rise and the howling ceases when this rate of temperature rise falls off.

As far as the mesh size is concerned, it is found that this affects the quality of the sound; a fine mesh produces a louder sound of longer duration than a coarse mesh. This of course is related to the temperature reached by the air just above the gauze.

Explanation of the effect

The hot gauze in the Rijke tube heats the air above it and as the hot air rises cold air is drawn in from below.

Lord Rayleigh's explanation of how this convection may be superimposed on the displacement of air particles due to a sound wave has been couched in modern terms by Ffowcs Williams and Dowling as follows.

If heat is added to a sound wave when it is in the high temperature stage of its cycle, energy is fed into the acoustic disturbance, but if heat is added during the low temperature stage the sound wave loses energy. The compressions and rarefactions in a sound wave are adiabatic and the pressure and temperature fluctuations are in phase, so that heat addition during a compression (positive pressure fluctuation) increases the amplitude of the sound wave, but has the opposite effect during a rarefaction (negative pressure fluctuation). Thus when the acoustic particle displacement is positive (upwards), cold air is drawn in from below the gauze to be heated up, but when the acoustic particle displacement is negative (downwards), relatively little heat transfer takes place. Hence we conclude that for maximum heat transfer to occur, the particle displacements and pressure fluctuations must be in phase. This occurs in the lower half of the tube. If the gauze were in the upper half of the tube, where the particle displacements are out of phase with the pressure fluctuations, there would be little heat transfer and the tube resonance would not be excited. This is in accordance with observation.

Conclusion

Students find this effect novel and interesting. There are many aspects that they can investigate, the amount of detail depending on the time available.

References

1 P.L.Rijke, *Phil. Mag.* **XVII** *(94th Series*, 419 (1859)
2 J.W.S. Rayleigh, 'On acoustical observations', *Phil. Mag. Vol. VII 5th series*, p.155-6 (1879)
3 J.W.S. Rayleigh *Theory of Sound*, p. 231-4, vol. 2 (1896)
4 J.E. Ffowcs Williams and A.P. Dowling, *Sound and Sources of Sound*, Ellis Horwood, p.128 (1983)

Ultrasound speedometer

P.J. McDonald and E.P. O'Reilly
Department of Physics, University of Surrey, Guildford, UK

Introduction

This experiment* is taken from our first year undergraduate laboratory. Most of this laboratory is based on extended mini-projects lasting five whole days (see our companion paper 'Depth Gauge'). We believe that this encourages the students to develop a wider range of skills than is the case in a traditional laboratory based on short set experiments.

In this project students are asked to investigate the properties of ultrasound and to evaluate its use as a speedometer. To this end they must measure speed in as many ways as possible and at the end of the project present a report evaluating which of the methods is most suitable for use as a speed trap. The best students can develop four or five different methods to measure speed whereas most students achieve at least two. The aims of the project are to teach wave phenomena such as Doppler shift and phase and to develop skills in practical electronics, experimental design, data analysis and report writing.

Apparatus

Initially the students are presented with a model train and a straight railway track, 6 m long. They are also given sufficient components from which to construct an ultrasound transmitter and receiver system and a light-independent resistor circuit. The light-dependent resistor can form the basis of a circuit capable of converting signals on an oscilloscope screen into electrical pulses. Use of this oscilloscope to pulse converter is described in more detail below. A light-dependent resistor (type ORP12) is advised as this has a sufficiently slow response not to be disturbed by screen refresh and the 50Hz mains lighting. A typical circuit for this converter is shown in Fig. 1. Suitable transducers and application

* Reprinted with permission from Physics Education

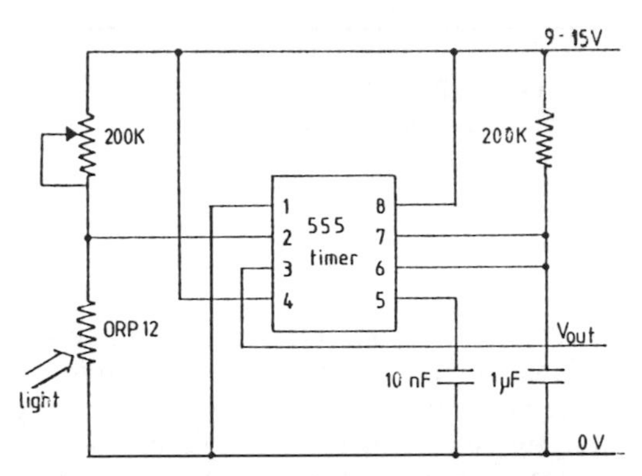

FIG.1 The experimental setup. Doppler shift measurements can be made using the two frequency meters (F). Phase shift measurements can be made by counting pulses, using the ORP12 circuit mounted on the CRO screen and the counter C

notes are readily and cheaply available from commercial suppliers. A block diagram of the assembled apparatus is shown in Fig. 2.

Experimental

(a) Doppler shift

The best known method of measuring speed is via the Doppler shift in frequency between the transmitted and received ultrasound waves. For this the transmitter and receiver can be mounted together so that the system works in reflection mode (it is then usually necessary for the train to carry a relatively large card to reflect the waves) or the transmitter can be mounted as a stationary observer with the receiver mounted on the train so that the system works in transmission mode. The Doppler shift in frequency in transmission mode for increasing transmitter - receiver separation is then given by

$$\Delta f = -\frac{v}{c_s} f$$

where Δf is the change in frequency f when the train is moving with velocity v and c_s is the speed of sound. Direct measurement of the Doppler shift is difficult using a single basic frequency meter. Fluctuations in frequency between two runs and instabilities in the meter can be as great as the frequency shift being measured,

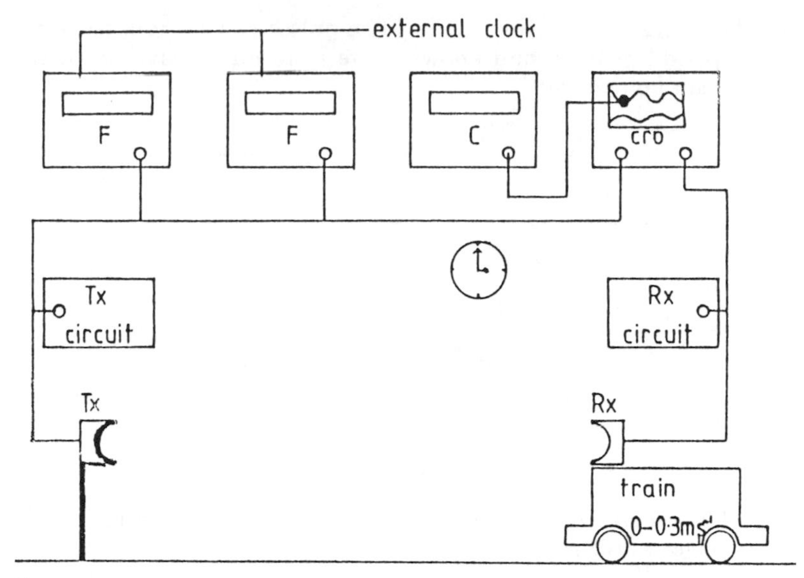

FIG.2 A possible circuit for producing electrical pulses from a switching light source. The 200 kΩ potentiometer sets the light threshold. The 200 kΩ resistor and 1 μF capacitor set the pulse length to about 0.2 s and may be changed for other pulse lengths

typically 5 Hz in 40 kHz. The measurement is easier if two counters triggered by a common oscillator are used simultaneously to monitor the transmitter and receiver, especially if the counters permit overflow of the (known) most significant figures. The counting period has to be timed to deduce frequency.

(b) Phase shift

The next set of methods involve measuring the phase shift between the transmitted and received waves. Because of the path difference between the transmitter and receiver, when the train moves through a single wavelength λ the phase of the received wave shifts by 2π in transmission mode and by 4π in reflection mode. The transmitter oscillator is used to trigger an oscilloscope and is displayed on channel 1. The received waveform is displayed on channel 2 and moves across the screen in sympathy with the train motion. The light dependent resistor is mounted on the oscilloscope screen so that the moving waveform peaks pass beneath it. The circuit generates a pulse each time a peak passes. Each pulse thus corresponds to a train displacement λ in transmission mode or $\lambda/2$ in reflection mode. The pulses can be counted for a fixed period of time, and knowing the ultrasound wavelength the speed can be deduced.

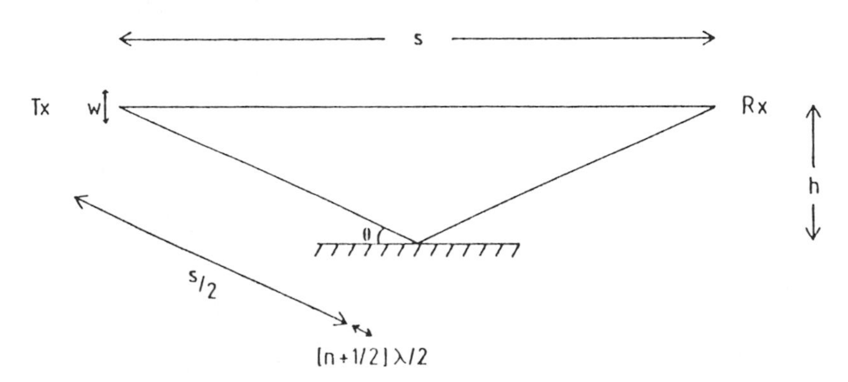

FIG.3 The geometry illustrating direct and reflected waves for destructive interference at the receiver

(c) Intensity

Finally the students can measure the decrease in received intensity with increasing distance from the transmitter. This method at first sight seems difficult to apply but in part due to good fortune works well and displays interesting physics. As well as the direct wave, there is a strong reflected wave from the bench top (see Fig. 3). The two waves interfere and in transmission mode this is easily detected. To calculate the positions of the intensity minima we first consider a point source and a point receiver at a height h above the bench. A sound wave striking the surface at an angle θ shifts in phase by 2θ on reflection. As will be seen below we are only interested in glancing angles, and we therefore assume zero phase change in the following analysis. Destructive interference then occurs when the direct and reflected waves differ in path length by $(n+\frac{1}{2})\lambda$, where n is an integer. From Fig. 3 this occurs when

$$(s/2)^2 + h^2 = \left(s + (n+\frac{1}{2})\lambda\right)^2/4 \qquad (1)$$

which on rearrangement gives the transmitter to received intensity minimum distance in the form

$$s = \frac{2h^2}{(n+\frac{1}{2})\lambda} - \frac{(n+\frac{1}{2})\lambda}{2} \qquad (2)$$

Successive minima or nodes are increasingly close to the transmitter, separated from each other by approximately $2h^2/(n^2\lambda)$.

In practice, the transmitter and receiver are not point sources. Let us consider the source to be distributed between heights h and $h+w$. The width of the nth minimum is then approximately

$$\frac{2(h+w)^2}{(n+\frac{1}{2})\lambda} - \frac{2h^2}{(n+\frac{1}{2})\lambda} \approx 4hw/n\lambda \tag{3}$$

Comparing (2) and (3), the ratio of the node width to node separation increases with node number n as $2nw/h$. The individual minima can therefore be detected for small n, that is for large receiver - transmitter separations. For larger n, the individual minima overlap too strongly to be clearly distinguished so that for small transmitter receiver separations a nearly constant finite intensity is observed. These effects are enhanced by the finite receiver size and the directional behaviour of the transducers. In our experimental setup with waves of $40\,kHz$, $h \approx 20\,cm$, and transducers 1 cm in diameter we can detect minima up to about $n = 5$, which occurs some 2 m from the transmitter. The speed can be deduced by timing the train between successive minima.

Conclusions

The transmission techniques can give the speed with a precision better than 10%. Reflection techniques are less accurate and reliable as they suffer more from spurious reflections from the laboratory. Overall, the project is suitable to students with a wide range of abilities, demanding a range of skills and ability in applied physics.

Ratio of heat capacities

P.H. BLIGH AND R. HAYWOOD
Kingston Polytechnic, Kingston upon Thames, Surrey, KT1 2EE, UK

Measurement as an art form

It is not what you measure as the way you measure it that matters. At least design technology wants us to enjoy the satisfaction and pleasure of a good performance. So every experiment requires careful stage management of hardware and software and must have a beginning, a middle, and an end.

And the experience has changed considerably over the centuries. The technology available to Reverend Stephen Hales, Vicar of Teddington, in 1733 when he became the first person to measure blood pressure, was rustic in nature. It consisted of a farmgate, the wind pipe of a goose, and some glass tubing. Hales strapped a mare to the gate and, with the help of his assistant, inserted a small glass tube into the artery of the neck(see Fig. 1). The flexible wind pipe of the goose acted as a piece of rubber tubing for joining the glass tube to a longer tube some 8 feet in length held vertically by his assistant. On releasing the clamp previously placed on the artery the blood rushed in almost to the top of the tube. The height fluctuated several centimeters with the beat of the mare's heart. Repeating the performance with the jugular vein, the blood rose only 20 inches. Hales went on to determine the amount of blood the heart pumps by bleeding the animal to death and filling the left ventricle of the heart with melted bees wax. Its volume on solidification was 160 millilitres, which with a heart rate of 36 beats per minute gives a pumping rate of six litres a minute – an underestimate as the mare's heart shrank after being bled.

Hales had the imagination and will to use the 'enabling technologies' of his time to make a dramatic measurement of blood flow. With similar ingenuity this gentle man measured plant growth with a success that led him to be called the father of plant physiology. A gifted experimenter, he turned his skill to improving the ventillation in prisons – it was he who suggested the method of collecting gases under water – and the winnowing of corn. More importantly it was he who invented the use of the tea cup to prevent the crusts of pies and tarts from collapsing!'[2]

FIG.1 Stephen Hales and his assistant measure the arterial pressure of a horse

We need such lateral thinkers to make experimental design imaginative and dramatic as in this measurement of pressure. Today the extensive market in novel sensors and transducers, many minaturised and integrated with their electronics, makes, for example, the automated non-invasive measurement of blood pressure a less painful process and more precise enterprise.[3] A modern instrumental system has the same functional components as in Hales bewitching senario, but it aims to minimise the human factor (Fig. 2). And the software that orchestrates the performance must evoke the right human response. This is the man-machine interface, which is all important and must be entertaining and insightful if the performance is to be worthwhile.

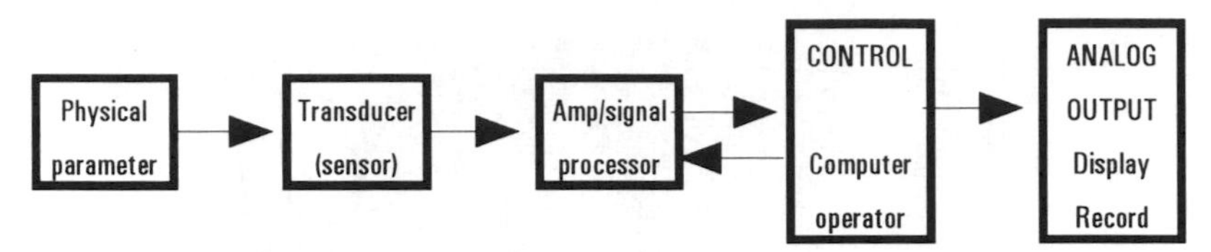

FIG.2 A diagram showing the function and components of a modern instrument

What's in an idea?

In 1929 Ruchhardt[4] had a clever idea for the measurement of the ratio of heat capacities (γ) that made nice use of some elementary mechanics and demonstrated some interesting physics in the process – as we shall see. His method was an improvement of that of Assmann[5] who in 1852 used the oscillations of a column of mercury enclosing a quantity of gas to study adiabatic changes in the gas.

Taking advantage of the fact that precision glass tubes and steel balls had become available in the market place, Ruchhardt found he could obtain balls that would fit so snuggly in the tubes as to be almost air tight and move like a piston with only a modest amount of friction. A diagram of his equipment is shown in Fig. 3. The gas was contained in a large jar into the neck of which was fitted the glass tube. A ball introduced into the top of the tube could perform 10 oscillations before coming to rest. If we assume the gas to be ideal and the oscillations to take place adiabatically, then gamma is related to the natural frequency of the oscillation, f_0, by,

$$\gamma = (4\pi^2 mV/A^2 P) f_0^2 \tag{1}$$

where P, V are the equilibrium pressure and volume of the gas, respectively, A is the cross sectional area of the metal ball (and glass tube), and m is its mass. To overcome the difficulty of measuring f_0, Rinkel[6] (1929) measured how far the ball fell when released from a position at which the gas was at atmospheric pressure P_0, a distance L given by,

$$L = 2mgV/P_0 A^2 \gamma \tag{2}$$

Not only the problem of measurement but the presence of friction (quite apart from the idealism of assuming the gas perfect and the oscillations completely adiabatic) was an inherent disad-

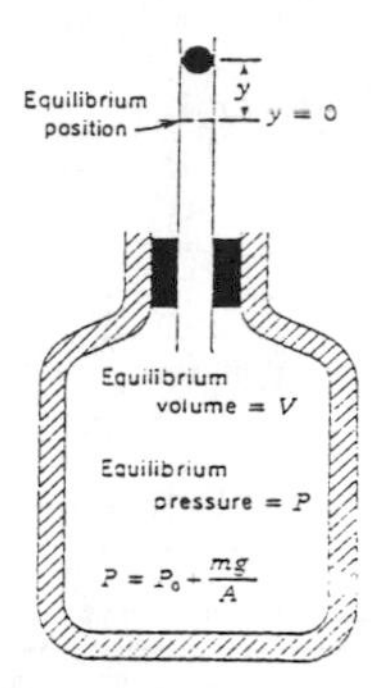

FIG.3 A diagram of Ruchardt's apparatus to measure the ratio of heat capacities

vantage. It was in fact the major source of error, about 3%, and depended on the degree of precision with which the ball fitted in the glass tube. In 1940 Clarke and Katz[7] modified the experiment to reduce the experimental difficulties and allow for the theoretical assumptions. In their version a steel piston is located at the centre of a cylinder containing the gas and is driven into oscillation by externally placed magnetic coils at any frequency as determined by the ac current tuned to determine the resonant frequency of the piston in the gas.

A lift magnet operating on the piston reduced friction to 1% (Fig. 4 is a diagram of the apparatus). Using the real equation of state and allowing for deviations from ideal adiabatic conditions this highly specialised equipment enabled gamma to be measured with considerable accuracy between 1 and 25 atmospheres. (The method was improved further by Katz, Wood and Leverton in 1949 enabling the resonant frequency to be measured to an accuracy of 1 part in a thousand[8]).

The problem with such equipment from an educational view point is that it must be specially developed and looses some of the simple elegance of the original operation and idea. J.L. Hunt (1985) recovered some of this for us when he noticed that in modern glass syringes, when clean and dry, the metal piston moves with as little friction as Richardt's metal balls.

In our version of Hunt's method a 50 cc glass syringe with metal plunger with Luerlock fitting was purchased from Orme Scientific Limited. A syringe stopcock (Aldrich Chemicals cat. no. 210, 325.0) was fitted to the outlet so that different gases could be introduced into the syringe from a football bladder and sealed in.

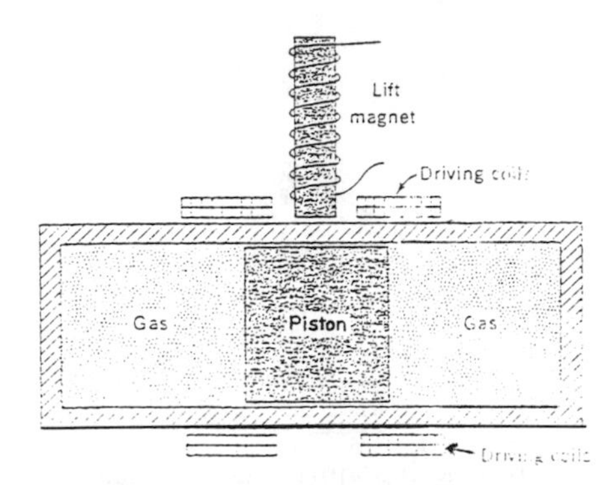

FIG.4 A diagram of the apparatus of Clarke and Katz from Zemansky (Ref. 6)

On compressing and releasing the plunger it can be seen to oscillate. The situation is visualised diagrammatically in Fig. 5 and should be compared with that of Ruchauldt shown in Fig. 3. It is a straight forward argument (see Appendix) to arrive at equation (1) for this apparatus also. The measured damped frequency f is related to the natural or fundamental frequency f_0 by the relation

$$f_0/f = [1 + (\lambda/2\pi)^2]^{1/2} \tag{3}$$

where λ is the natural logorithm of the ratio of successive amplitudes of the oscillation, i.e., from Fig. 5:

$$\lambda = \ln(A_1/A_2) \tag{4}$$

where λ is called the logarithmic decrement of the oscillation.

Enter the new technology

The advent of relatively cheap transducers often with integral electronics has made possible the automatic monitoring of physical parameters – in this case an accelerometer to monitor mechanical vibrations. The display and measurement of such oscillations no longer depends on the possession of an expensive storage oscilloscope. A sufficiently fast A/D converter coupled to a cheap microcomputer can, with some modest software, display the oscil-

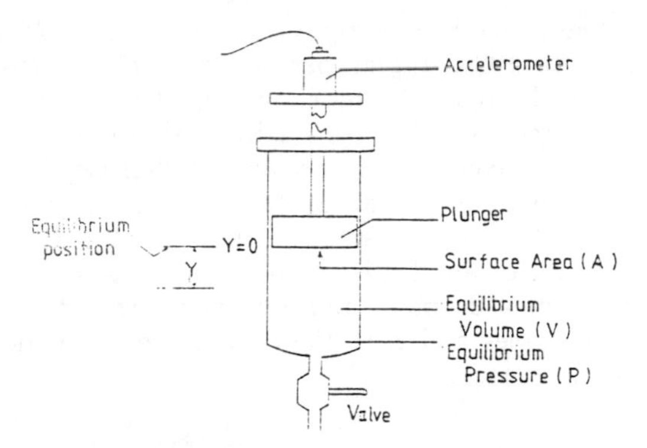

FIG.5 A diagram of the glass syringe used to measure the ratio of heat capacities and a visualisation of the oscillation of the plunger

lation, measure its frequency and damping, and calculate the heat capacity ratio.

A piezoelectric charged coupled accelerometer with integral electronics was fitted to the plunger system.[10] When the crystal is accelerated it experiences a force F that distorts it, causing a small displacement x of atoms in the crystal from their normal position, such that

$$x = \frac{1}{k}F \tag{5}$$

where k is the stiffness of the crystal, which is large and typically $2 \times 10\mathrm{N\,m^{-1}}$. This deformation of the crystal lattice results in a net charge displacement q proportional to x so that

$$q = bx \tag{6}$$

where b is a constant. Combining (5) and (6) gives

$$q = \frac{b}{k}F = hF \tag{7}$$

where $h = b/k$ coulombs per newton (or coulombs per gram) is the charge sensitivity to the force. The charge is then picked up by the two electrodes and fed into a minaturised hybrid circuit (a.c. coupled with a constant current source) integrated into the device which converts the high impedance pC/g signal to a low impedance mV/g signal. This signal is then fed into an all purpose variable gain amplifier designed to work with a variety of sensors (e.g., strain gauges).

The amplified signal is now digitised using a fast A/D converter, working at up to 100 samples per second. The continually updated signal can be read off the converter at a speed required by the microcomputer. In this case a BBC microcomputer was used whose own A/D converter was too slow to digitise the signal and lacks the reproducibility to accurately measure signal frequency and amplitude. The signal was fed into the 1 MHz bus of the BBC microcomputer for display and analysis. A flow diagram of the acceleration measurement and data analysis system is given in Fig. 6. Figure 7 gives a circuit diagram of the box containing the amplifier and converter.

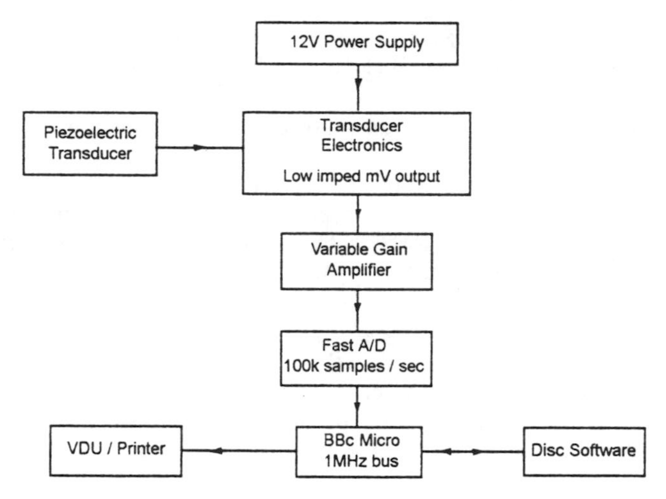

FIG.6 A flow diagram of the acceleration measurement and data analysis system

Operational design

Software has been designed to display and analyse the plunger oscillation and from it calculate the heat capacity ratio for the gas in the syringe.

The experiment consists in filling the syringe with the gas from the bladder to an appropriate volume and closing the stop cock.

Using OPTION 1 on the software the computer is turned into a voltmeter with 10 mV precision with an attractive display scale. In this instance it is used to zero the amplifier output (see Fig. 8).

See figure 7 on page 99.

Fig.7 Circuit diagram for the BBC IMHz bus Analogue Interface Unit which contains a general purpose amplifier, fast A/D converter, eight additional A/D converters and one A/D converter (further details supplied on request)

OPTION 2 is now chosen to display the oscillation itself and analyse it. The plunger is depressed a few cm^3 and then released whilst simultaneously prompting the computer to receive data. An example of the consequent display that can be obtained, in this case with nitrogen, is shown in Fig. 9. The software enables the experimenter to operate two cursors that can be moved along the oscillation, and their positions in terms of amplitude (in volts) and time (in seconds) are simultaneously recorded below a display of the oscillation.

OPTION 3 of the software enables the experimenter to feed in the necessary data as requested and calculates the ratio of heat capacities for the gas.

The approach to measurement offered here may take much of the spice out of experimentation as enjoyed by the Reverend Stephen Hales. The required skills have changed but need to be no less dramatic. We must control our primitive instincts to do it all ourselves, swing our own pendulums and measure their motion with our heart beat. There has to be a greater willingness to put our hands in our pockets and spend more time thinking imaginately about designing the performance, programming its operation, and

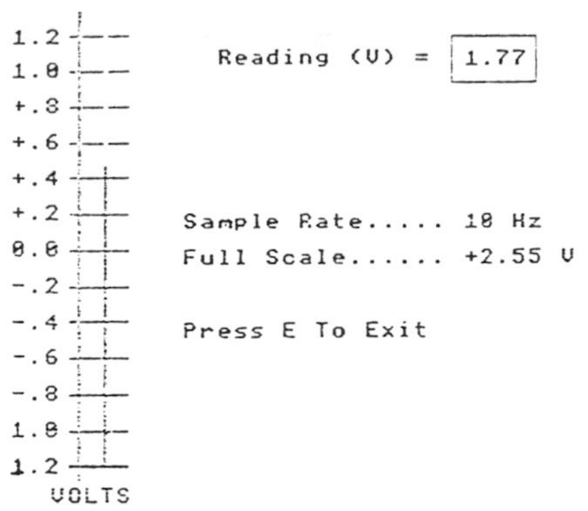

1.2
1.0
+.8
+.6
+.4
+.2
0.0
-.2
-.4
-.6
-.8
1.0
1.2
VOLTS

Reading (V) = 1.77

Sample Rate..... 10 Hz
Full Scale...... +2.55 V

Press E To Exit

Fig.8 The display scale on the microcomputer when used as a voltmeter.

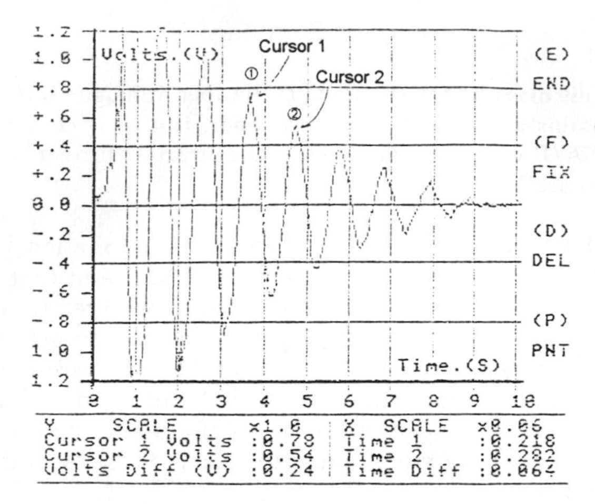

FIG.9 Computer display of plunger oscillations and analysis of its amplitude and frequency.

displaying its achievement. Then we can sit back and enjoy the performance as should any true artist and share his or her pleasure in communicating that enjoyment to others.

Appendix: Derivation of the equation for the ratio of heat capacities of a gas by the glass syringe method

With reference to Fig. 5, the pressure of the gas enclosed in the syringe is

$$P = P_0 + \frac{mg}{A} \tag{A1}$$

where

P_0 = atmospheric pressure,

m = mass of plunger system (including accelerometer),

A = area of the face of the plunger,

g = acceleration due to gravity.

If the plunger is depressed a small distance y as shown, the pressure in the gas will increase by a small amount ΔP so that it exerts an equal and opposite restoring force on the plunger of $A\Delta P$. The corresponding decrease in volume is $\Delta V = yA$.

Since the changes in P and V take place rapidly they are assumed to be adiabatic. As the changes are small we assume they are also quasi-static. If we also assume the behavior of the gas is such that it obeys the perfect gas laws, then the changes in P and V during the oscillations are described by the equation

$$PV^{\gamma} = \text{constant} \tag{A2}$$

where γ is the ratio of the heat capacities of the gas at constant pressure (C_p) and constant volume (C_v) i.e.,

$$\gamma = C_p/C_v \tag{A3}$$

We now differentiate equation (A2):

$$P\gamma V^{\gamma-1}dV + V\gamma dP = 0 \tag{A4}$$

which for small finite changes ΔP and ΔV in P and V during the oscillation gives to a good approximation

$$P\gamma\Delta V^{\gamma-1}V + V^{\gamma}\Delta P = 0 \tag{A5}$$

taking the appropriate substitutions the restoring force F is given by

$$F = A\Delta P = -(\gamma PA^2/V)y \tag{A6}$$

As the restoring force is directly proportional to the displacement y, this is the condition for the plunger to experience simple harmonic motion (see Fig. 9 for confirmation) such that

$$F = ma = -ky \tag{A7}$$

where a is the acceleration experienced by plunger system of mass m and $k = \gamma PA^2/V$ a constant of proportionality describing the nature of the restoring force.

Equation (7) can be rewritten as

$$a = \frac{d^2y}{dt^2} = -\frac{k}{m}y \tag{A8}$$

a solution of which is $y = A\cos w_0 t$ where

$$w_0 = 2\pi f_0 = \sqrt{\frac{k}{m}} \tag{A9}$$

and f_0 is the natural frequency of oscillation of the system.
This may be rewritten to give

$$f_0 = \frac{1}{2\pi}\sqrt{\frac{\gamma P A^2}{mV}} \tag{A10}$$

or

$$\gamma = (4\pi^2 mV/A^2 P) f_0^2 \tag{A11}$$

i.e., equation (1) in the text.

References

1 The picture is by the late John Fulton and reproduced in: Eric Neil, *William Harvey and the Circulation of the Blood*, Priory Press, p.89 (1975)
2 Charles Singer, *A Short History of Biology*, O.U.P. 364 (1931)
3 See the buyer's guide to transducers in *Transducer Technology*, June 1987
4 E. Ruchardt, *Phys. Z.* **30**, 58-59 (1929)
5 A. Assmann, *Ann. Phys. Z.* **85**, 1 (1852). Referred to in J.K. Roberts and A.R. Miller, *Heat and Thermodynamics*, Blackie and Son, 5th ed. (1960)
6 Referred to by M.W. Zemansky, *Heat and Thermodynamics*, McGraw Hill, 5th ed., 130 (1968)
7 A.L. Clarke and L. Katz, *Can. J. Res. A* **18**, 23-28, 39-63 (1940); **19**, 111, (1941); **21**, 1 (1943)
8 L. Katz , S.B. Wood, and W.F. Leverton, *Can. J. Res. A***27**, 27-38 (1949)
9 Vibro-meter Corp. model CE508 M201. See Ref.3 for alternative systems

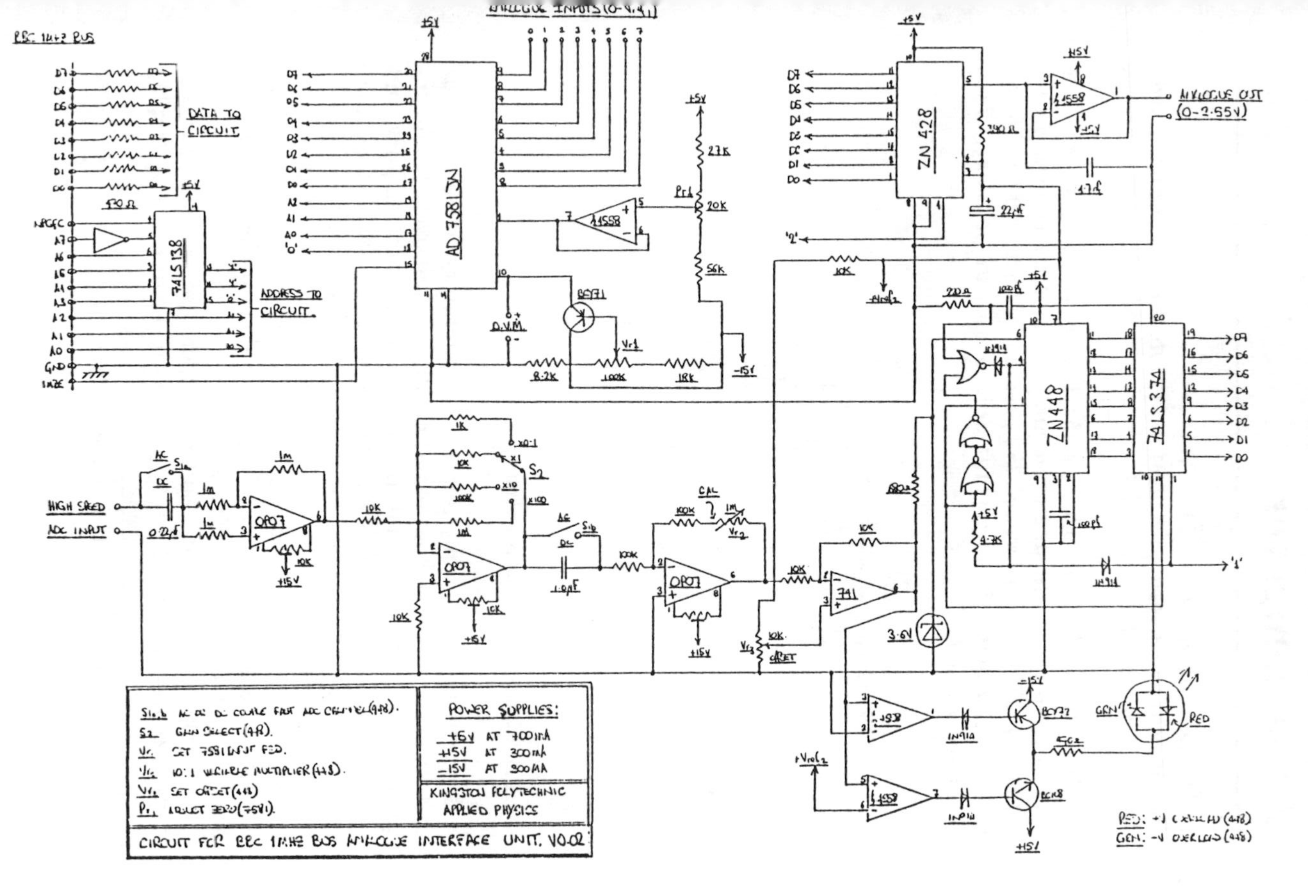
DATA TO CIRCUIT
ADDRESS TO CIRCUIT
74LS138
AD 7581JN
ZN 428
ZN 448
74LS374
ANALOGUE OUT
(0-2.55V)
HIGH SPEED
ADC INPUT
POWER SUPPLIES:
+5V AT 700mA
+15V AT 300mA
-15V AT 300mA
KINGSTON POLYTECHNIC
APPLIED PHYSICS

Heat capacity of solids by courtesy of the computer

P.H. Bligh and R. Haywood*

Department of Physics, Kingston Polytechnic, Kingston upon Thames, Surrey, KT1 2EE, UK

'Hello! We are looking today at the heat capacity of solids. We will make some simple measurements and investigate their implications. I will lead you through the program but you will have to set the pace and make the important decisions as we go along. Type OK when ready to go'.

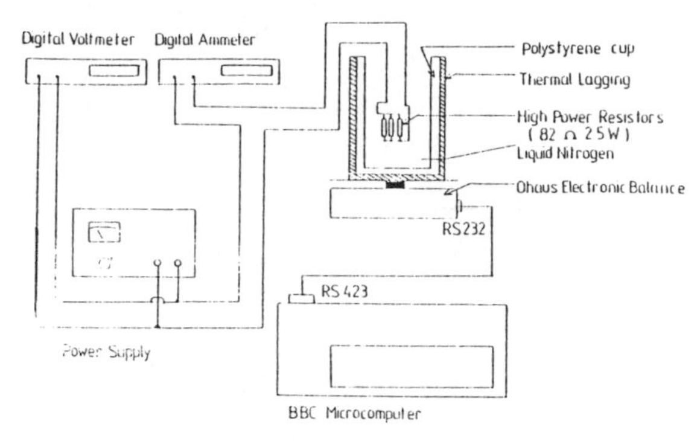

Fig.1 Block diagram of apparatus

The arrival of the computer in the laboratory has enabled us to turn some rather trivial experiments into relatively useful investigations with a pleasant mixture of theoretical enquiry and data manipulation. Time is the laboratory traffic warden restricting student parking on one particular experiment to two to three hours. The crucial question so often is, how quickly and effectively can we get the student into the driver's seat of an experiment so thatthe student's enquiry can be guided along preferred routes to a destination of physical interest?

Our (or should we say student) experience is that practical physics consumes time in bulk, what with the drawing of graphs,

* Reprinted with permission from Physics Education

the calculations, estimations of significance and errors, the all-too-thin critical discussion and comparison of results with standard data or theoretical models. All too often there are not enough small hours in the week or books in the library or staff available to reach a satisfactory intellectual appreciation of the actual deed done in the laboratory.

In what follows we see how the computer - plus its software - can come to our (or rather the student's) rescue. Rather than nail the student to a laboratory bench and compel him or her to follow a set of instructions that generate a forest of graphs and a jungle of data, it enables the student to use these as menus rather than hard-won ends to achieve a far more impressive experience of the subject at hand.

The experiment considered here is one that has been at work in our laboratories for several years. One form of it has recently been described in the literature.[1] In principle, it is the old calorimetric experiment of junior physics days measuring heat lost against heat gained. But when we discover, for example, that the modern osmometer works on corresponding elementary physics principles at a relatively high cost, we realise that what matters most is how well these elementary principles of physics are applied to solve each particular problem. In our particular experiment about 25 g of liquid nitrogen (LN2) in a polystyrene cup (from the refectory, lagged to prevent water freezing on its outer surface) is placed on a top loading balance together with a few grams of metal whose heat capacity is to be determined. A small piece of thread is attached to the metal. Fourtunately, a top-loading balance (Ohaus Electronic RS 232) is on the market at a few hundred dollars and can interface directly with a microcomputer. The computer can reset the balance to zero and read directly the load on the balance as instructed. In our case a BBC microcomputer was used. The layout of the experimental setup used is shown in Fig. 2.

The heat capacity experiment is designed to provide an interactive environment between the student, the experiment and the computer. Software has been written so that the whole event can be orchestrated from the computer keyboard.

The procedure is stored in several program overlays: a general purpose program for collecting data from the Ohaus balance and storing it on disk; an analysis package for plotting data, solving linear regression, and printing the results; tables of standard data and interrogative questions. It also includes a versatile data plotting program DATPLOT written by Alan Vincent[2] that will label and scale axes once chosen and make a regression fit to the data if

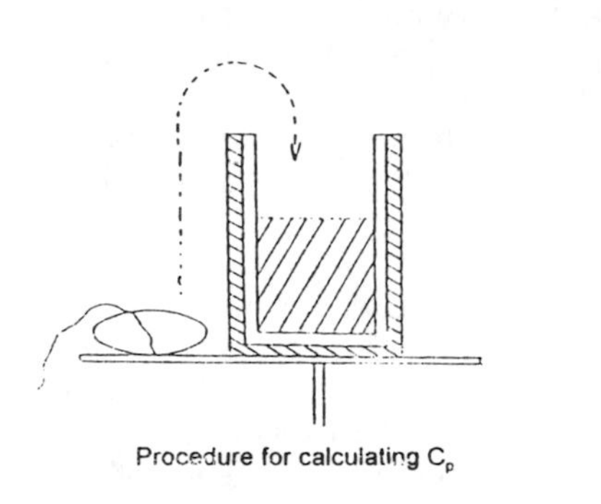

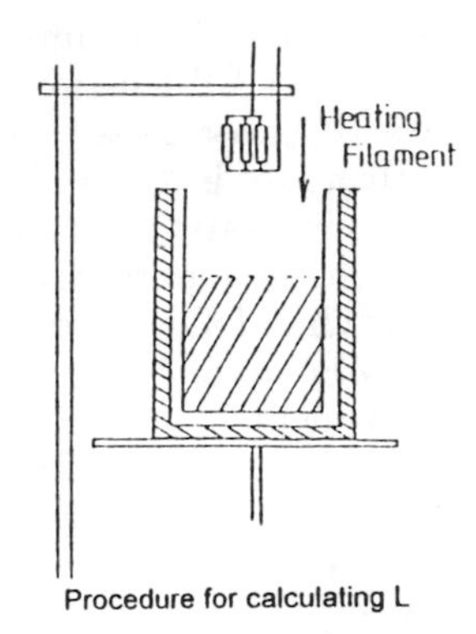

FIG.2 Layout of the experimental setup

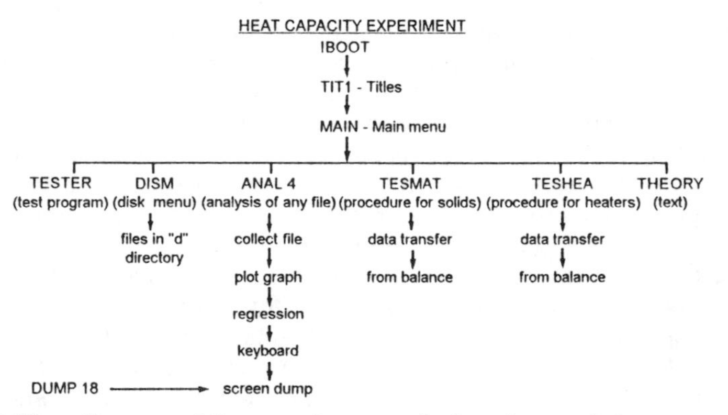

FIG.3 Flow diagram of the experiment as designed around the computer software

necessary.

These programs are based around a MENU program MAIN, which formulates the design of the whole experiment. A flow diagram is given in Fig. 3.

Although the programs and the accompanying experimental notes aim to guide it is necessary for the student to decide WHEN and HOW to use each program.

At the very outset the student can call up the experimental notes as given in Fig. 4 as a basis for exploration. The student can then 'play with the programs' – calling up standard data relating to the materials to be studied, viewing some test data, looking at the interrogative questions, so that a process of *familiarisation* takes

```
 120CLS:PRINT''"Place beaker AND sample on BALANCE ready to tare...":PROCstart:
PROCtare
 130CLS:PRINT''"Add LN2 (Approx half beaker) CAREFULLY....":PROCstart
 140CLS:PRINT'' "When ready to begin recording...."':PROCstart
 150CLS
 160ENDPROC
 170DEFPROCtext
 180PRINT'"Experimental Procedure"''"CAREFUL - LN2 is a hazardous substance."'"
Glasses and gloves MUST be worn for all manipulations....."
 190PROCtime10:PRINT'"...Apart from pressing MY KEYS!":PROCtime10
 200PRINT'"First obtain a heat loss curve for LN2 to the environment. How woul
 you do this? Press SPACE BAR for answer."
 210 REPEAT:UNTIL GET=&20
 220PRINT'"TARE the balance as instructed."'"Add about 25g LN2 to the plastic
p. "'"The computer will take readings of"'"weight and time elapsed about ever
 en"'"seconds"'"After about 5 minutes, lift the weight"'"by the thread";
 230PRINT" and SLOWLY lower the"'"sample into the LN2. Continue with"'"readinc
 until the graph is parallel to"'"the original heat loss curve."
 240PROCtime10
 250PRINT'"Are you ready to GO?":REPEAT:G=GET:UNTIL G=&59 OR G=&79

 330VDU5:MOVE 0,50:PRINT"SPACE BAR to end":MOVE 600,50:PRINT"G for GRAPH"
 332MOVE 600,90:PRINT"L increases readings"
 335MOVE 900,1000:PRINT"CAREFULLY"
 336MOVE 900,950:PRINT"add sample"
 337MOVE 900,900:PRINT"after about"
 338MOVE 900,850:PRINT"200 secs"
 340MOVE 50,1000:PRINT"Weight(g)","  time(secs)":VDU4
```

FIG.4 Experimental notes as presented to the student initially

place.

The student then runs the experiment to measure the latent heat of liquid nitrogen (LN2) and then uses this result to determine the specific heat (C_p) of four different metals. This experiment consists basically of lowering the metal at room temperature (T_0) into LN2 at its boiling point of 77 K and determining C_p from the amount of LN2 that evaporates. The value of C_p obtained is taken to be the average heat capacity in the temperature range 77 K to T_0. Details of these two experiments are as follows.

Led along by the software a plot of the mass of LN2 in the cup as a function of time is obtained for about 5 min. The student is then instructed to immerse the heater (high power resistors) into the LN2 for a time Δt at a voltage V and current I. On removal of the heater the computer continues to plot the mass loss curve of the LN2 for a further five minutes. Using the cursor on the displayed graph the student may decide the starting and finishing point of each cooling curve and then the computer makes a regression fit to each. Either by eye from the graph or by using the regression equation the student measures the vertical distance between the two regression lines, half way through the heating process. This will give the student the mass loss of LN2 due to evaporation (Δm). Then the latent heat of nitrogen (L) is obtained from the expression

$$IV\Delta t = \Delta mL \tag{1}$$

The student types the appropriate values into the computer as requested and the final results and calculations are presented as in Fig. 5.

```
Final Calculation Of the
Latent Heat of Vaporisation of LN2

Type in mass lost (grams) Δm?8.6

Type in the Heating time (seconds) Δt?12
0

HEATER Voltage was 20.00 volts

HEATER Current was 0.65 amps

      I.V.Δt = Δm . L

HENCE : L = 182.51 J per gram

Standard Value: L = 197.2 J per gram

Press SPACE BAR to return
```

Fig.5 Presentation of data and result for latent heat of vaporisation of nitrogen (LN2)

Having obtained L the experiment is repeated but this time heat is supplied by lowering the solid into the LN2 (Fig. 2). Once again the student is led along by the software.

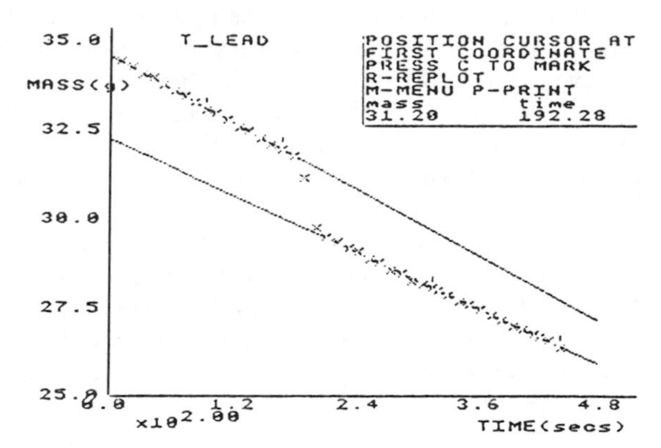

Fig.6 Plot of data for weight of cup plus metal and LN2 as a function of time

Figure 6 is an example print-out of the weight of the cup plus metal and LN2 as a function of time obtained for lead with regression fits made to the two halves of the curve before and after the lead was added. Fig. 7 is the presentation given of the final data and calculated specific heat capacity for lead using the expression

$$M\bar{C}_p(T_0 - 77) = \Delta m L \tag{2}$$

where M is the mass of the metal used and $\bar{C}_p$ is the average value for C_p between room temperature T_0 and 77 K.

```
Final Calculation of the
Heat Capacity of PB

Latent Heat of LN2 = 197.7 J/g

Atomic mass of sample = 207.20

Mass of sample = 14.40

HENCE :- 0.07 moles of sample

Initial temperature = 299.70

Final temperature = -77K

Type in mass of LN2 lost <grams> 22.1

HENCE :-  C̄p = 26.82 J / mole K

Press SPACE BAR to return
```

FIG.7 Presentation of data and result for the specific heat of lead.

The experiment is performed for four different materials: beryllium (Be), carbon (C), silicon (Si), and lead (Pb). This would present an impossible task for a student to complete in three hours if all the associated graph work and calculations had to be completed as well. Neatly and effectively, the computer presents the student with data files of the results for each metal.

At this point we have reached the beginning rather than the end of the experiment as far as the real physical appreciation of the results is concerned. So having arrived at this point the student refers to the question file -CHAIN 'QUEST' - to investigate the subject of specific heats still further.

'From tabulated data calculate the average specific heat capacity for each metal in the range 77 K to T_0 by numerical integration. How do these compare with your experimental values?' The student displays the standard values of specific heat for the four metals over the temperature range 77 K to 300 K using CHAIN TABLE 1 (Fig. 10) and, using a numerical integration of the form

$$\bar{C}_p = \sum C_p \Delta T / \sum \Delta T \tag{3}$$

calculates $\bar{C}_p$ over the range 77 K to T_0 from these data.

Another program in the menu (THEORY) confronts the student with a theoretical model for *deriving* heat capacities in solids - the Debye model - and asks the student to consider the results

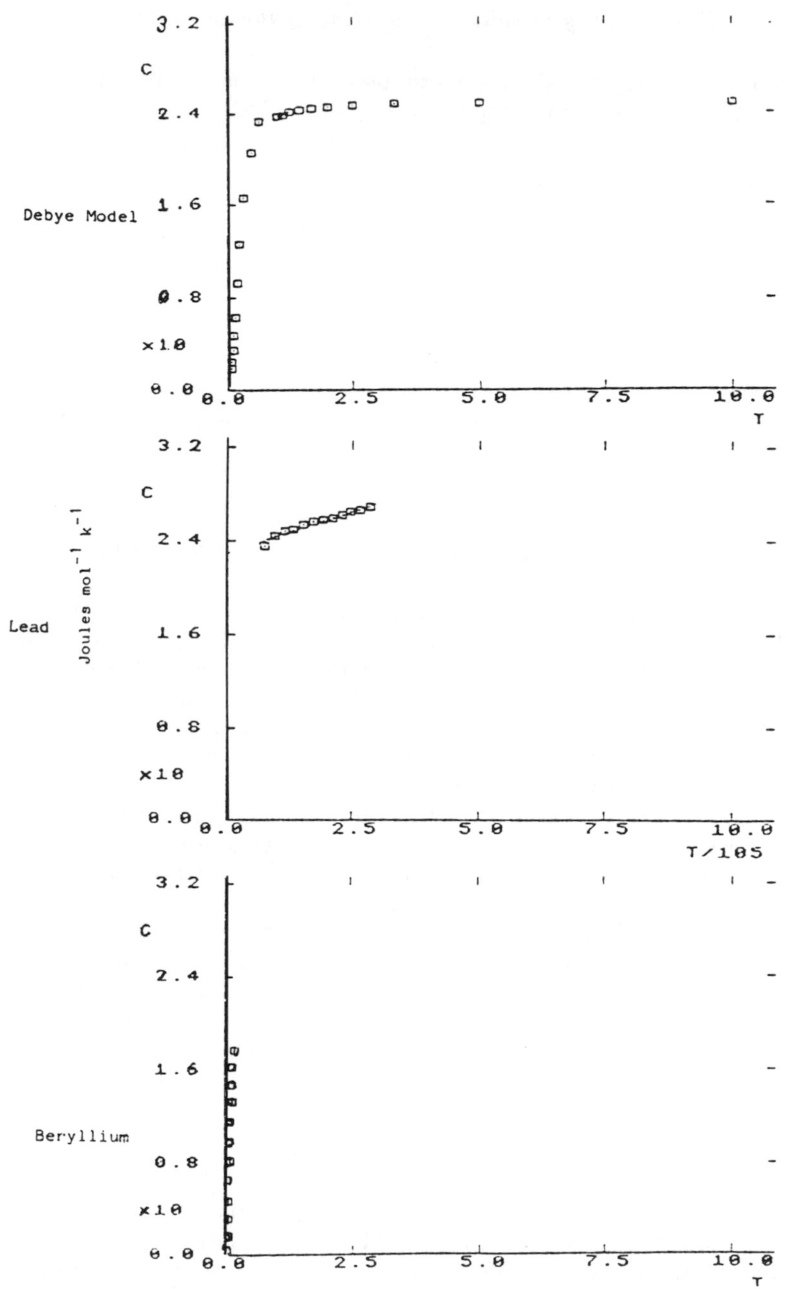

FIG.8 Plot of specific heat capacity (C) as a function of (T/θ) for the Debye model, lead, and beryllium in the range 80 to 300K

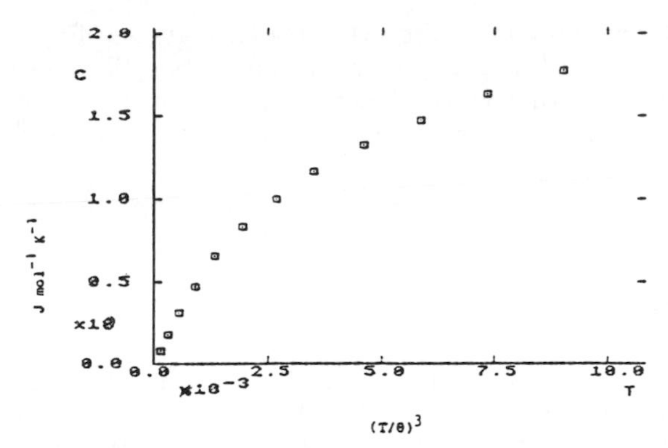

FIG.9 Plot of specific heat capacity (C) as a function of $(T/\theta)^3$ for beryllium in the range 80 to 300 K

alongside the model's predictions. Attention is drawn to the fact that, in the model, each metal has a characteristic temperature θ, the Debye temperature, at which the specific heat capacity at constant volume C_v reaches about 95% of its maximum classical value of 3R (where $T \gg \theta$). On this model when $T/\theta < 0.1$, we can make the approximation,

$$C_v(\text{Debye}) = (1945\,\text{J}\text{mol}^{-1}\text{K}^{-1})(T/\theta)^3 \tag{4}$$

which is the so-called T^3 law.[3] 'But *you* have measured C_p *not* C_v!' – chirps the computer piously – 'How are they related?' The appropriate prompt gives the answer that the difference of the two molar heat capacities is

$$C_p - C_v = TV\alpha^2 K \tag{5}$$

where α is the thermal expansivity, K the bulk modulus for the material and V its molar volume.[4] 'Is this difference significant in our experiment? CHAIN TABLE 2 (Fig. 11) to help you decide.' The data presentation is informative and makes for an easy answer.

The student is now encouraged to use DATPLOT and the information from TABLE 1 to plot values of C_p against (T/θ) for the different metals and compare these graphs with that for the Debye model, using data held in FILE 'DEBYE'. A comparison of such graphs for lead and beryllium with the Debye model in the range of temperatures we are considering shows that we are looking at lead in its classical region and beryllium in the quantum-mechanical part of the overall curve (Fig. 8). Finally, a graph of specific heat capacity against $(T/\theta)^3$ may be obtained for beryllium as shown in Fig. 9 to confirm that we are watching its quantum rather than its classical behavior in that equation (4) holds up to $(T/\theta)^3 \sim 10^{-3}$.

VALUES OF Cp - MOLAR HEAT CAPACITIES
AT CONSTANT PRESSURE IN J mol-1 K-1

TEMP	BE	C	SI	CU	PB
80	0.816	1.163	5.19	13.02	23.6
100	1.791	1.682	7.2	16.15	24.44
120	3.109	2.256	9.08	18.29	24.86
140	4.73	2.883	10.83	19.88	25.07
160	6.52	3.553	12.51	21.09	25.49
180	8.3	4.265	14.06	21.97	25.69
200	10	4.972	15.4	22.6	25.9
220	11.63	5.69	16.53	23.14	26
240	13.27	6.44	17.37	23.56	26.32
260	14.77	7.16	18.12	23.9	26.53
280	16.32	7.87	18.83	24.19	26.7
300	17.74	8.58	19.46	24.52	26.95

FIG.10 Values of molar heat capacities at constant pressure in $\mathrm{J\,mol^{-1}\,K^{-1}}$

If this were not enough, the computer finally presents the student with a summary series of questions to which, with the information by then available, the student should be able to go away and formulate answers.

The approach is certainly *not* the right one for all experimental work! Far from it. The student does not get his or her hands dirty. Indeed the means and ends of the program are not of the student's own making. Motor activity and measurement are minimal. But the procedure is participatory, provoking, and above all interesting. The student becomes involved in the subject, not via the library book, but the laboratory bench, through a friendly machine that offers to make the subject as pleasant in the laboratory as the hot drinks machine does during the half-way break in the refectory.

	BE	C	SI	CU	PB
At.weight	9.01	12	28.09	63.54	207.2
Debye Temp (K)	1440	2230	645	343	105
Bulk Mod(Nm-2) (x10+10)	12	3.3	10	13.8	4.58
Mod.Therm. Exp.(75K)	1.5	14.7	-1.5	22.8	73.2
(K-1(x10-6)) (293K)	33.6	23.4	7.5	50.1	86.1
Molar Vol m3(x10-6)	4.87	5.3	11.6	7.11	18.27
Cp-Cv J/mol.K	3.27E-2	1.19E-2	2.01E-3	0.239	0.978

FIG.11 Standard data for some materials

References

1 C.W. Thompson and H.W. White, *Am. J. Phys* **51**, (4), 362-364 (1983)
2 A. Vincent, *Education in Chemistry* **22**, (1), 22-23 (1985)
3 C. Kittle and H. Kroemer, *Thermal Physics*, W.H. Freeman (1980)
4 S. Glasstone, *Textbook of Physical Chemistry*, Macmillan (1956)

Latent heat – its meaning and measurement*

P.H. Bligh and R. Haywood

School of Applied Physics, Kingston Polytechnic, Kingston upon Thames, Surrey, KT1 2EE, UK

Abstract: **An elementary formalism is developed for latent heat so as to demonstrate its physical significance and to obtain expressions by which its value may be determined and its properties studied. Software has been written to enable students to investigate these properties in a flexible manner, using a standard experimental setup. Data files in the menu enable standard data to be called upon for analysis in a comparable way to experimental data. The use of transducers helps students to appreciate how thermodynamic parameters can be automatically monitored during a change of state.**

Elementary formalism for latent heat

Throughout this elementary formal treatment of latent heat we shall avoid all use of partial differentials and consider small incremental changes in thermodynamic parameters. Also, unless otherwise stated, the parameters used will refer to *molar* quantities of the substances considered.

What is Latent Heat? Following Glasstone,[1] an intuitively satisfying approach to the consideration of changes of state is to start by assuming that the Boltzmann distribution holds true for a liquid in equilibrium with its (saturated) vapor, i.e.,

$$n_V/n_L = \exp(-L_i/RT) \tag{1}$$

where n_V and n_L represent,respectively, the number of moles per unit volume in vapor and liquid at temperature T and L_i is the corresponding mean potential energy difference per mole between molecules in their vapor and liquid states. Equation (1) shows us that, because n_L varies little with T, increasing T increases n_V, so that the saturation vapor pressure must increase with temperature (Fig. 1). So L_i, sometimes called the *internal latent heat* of

* **Reprinted with permission from the European Journal of Physics**

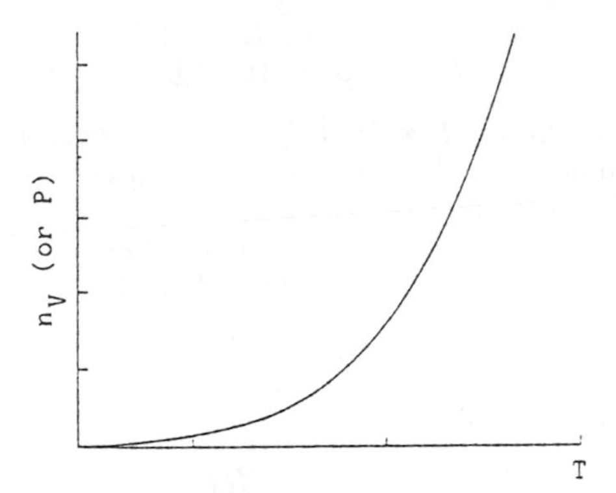

FIG.1 Sketch of logarithmic relation between concentration of molecules in vapor (n_V) or vapor pressure (P) and absolute temperature (T)

vaporisation, is equal to the difference in *internal energy* per mole (ΔU) between vapor and liquid.

The first law of thermodynamics states generally for any system that when an amount of work W is done *on* a system and an amount of heat Q is added *to* the system, then the increase in its initial energy is given by

$$\Delta U = Q + W \tag{2}$$

If we consider systems where W is entirely due to a small change ΔV in volume caused by the application of a pressure P, then $\Delta W = -P\Delta V$ and equation (2) becomes

$$\Delta U = Q - P\Delta V \tag{3}$$

If we suppose that the entry of heat Q into the system simply converts one mole of liquid into one mole of vapor *without* a change in volume then from (3)

$$\Delta U = Q = L_i \tag{4}$$

Normally expansion occurs during vaporisation *against* an external pressure P so that

$$W = -P(V_V - V_L)$$

where V_V and V_L are the molar volumes of the vapor and liquid at the temperature (T) and pressure (P) at which vaporisation occurs. In this case

$$L = Q = \Delta U + P(V_V - V_L)$$

or

$$L = L_i + P(V_V - V_L) \tag{5}$$

L is defined as the ***total molar latent heat or enthalpy of vaporisation*** and is equal to the heat required to evaporate 1 mole of liquid at constant pressure and temperature.

Usually $V_V \gg V_L$ and if we assume the vapor behaves approximately as an ideal gas then for one mole of the vapor,

$$PV = RT \tag{6}$$

and so (5) may be written,

$$L = L_i + RT \tag{7}$$

What is the relation between latent heat and specific heat? Suppose that, at temperature T_1, the latent heat is L_1 which is equal to the enthalpy difference per mole ΔH_1 between the vapor and liquid states at T_1. A small rise in temperature ΔT will result in a small change in L_1 to give a new value

$$L = \Delta H_1 + (C_V \Delta T - C_L \Delta T) \tag{8}$$

where C_V and C_L are the molar specific heats of vapor and liquid phases at constant temperature T_1. In general we may express this in integral form to give the latent heat at any temperature T_2 so that,

$$L_2 = L_1 + \int_{T_1}^{T_2} \Delta C_p dT \tag{9}$$

where $\Delta C_p = C_V - C_L$ at a given temperature T.

Figure 2 illustrates the relationship between the latent heat of vaporisation for water and temperature and its divergence from linearity.

Over modest temperature ranges well away from the critical temperature (647.2 K or 374°C for water) ΔC_p can be taken as approximately constant and we may write L as a linear function of temperature, i.e.,

$$L_2 = L_1 + \Delta C_p(T_2 - T_1) \tag{10}$$

If we take water as an example, between 0°C and 100°C with $C_L = 75.24\,\mathrm{JK^{-1}mol^{-1}}$ and $C_V = 36.00\,\mathrm{JK^{-1}mol^{-1}}$ then $\Delta C_p = -29.24\,\mathrm{JK^{-1}mol^{-1}}$, so that using equation (10),

$$L_T = L_{273} + (-29.24)(T - 273) \tag{11}$$

At $T = 373\,\mathrm{K}$, this equation gives $L = 42.09\,\mathrm{JK^{-1}mol^{-1}}$ which compares well with the published value of $40.62\,\mathrm{JK^{-1}mol^{-1}}$.

We should expect ΔC_p to be *negative* because L must decrease toward the critical temperature (T_c). At T_c the two phases become indistinguishable in that no heat energy is required to make the transition. The densities and heat capacities of both phases are therefore the same, i.e., $\Delta V = 0\,(n_V = n_L)$ and $\Delta C_p = 0$, so that $L_c = 0$. As ΔC_p approaches zero it will become very sensitive to changes in C_V which in general increases linearly with temperature.

How are vapor pressure, temperature and latent heat related? If we differentiate equation (1), we obtain,

$$\frac{d(\ln n_V)}{dT} = \frac{L_i}{RT^2} \tag{12}$$

We have assumed that n_L is independent of T as compared to n_V. If we further assume that our vapor obeys the ideal gas equation, we find from equation (6) that $P = n_V RT$ which gives

$$\ln n_V = \ln P - \ln RT \tag{13}$$

If we differentiate this expression with respect to T and compare our result with (12), we see that

$$\frac{d(\ln P)}{dT} = \frac{L_i + RT}{RT^2} = \frac{L}{RT^2} \tag{14}$$

This expression demonstrates the same logarithmic relation between vapor pressure and temperature as we have seen between concentration and temperature illustrated in Fig. 1. This is one form of the famous Clapeyron-Clausius equation, showing how L varies with P and T. If we again use the ideal gas equation $PV_V = RT$, where V_V is the molar volume of the vapor, we can express it in more familiar terms as

$$\frac{dP}{dT} = \frac{L}{TV_V} \tag{15}$$

Equation (14) is more useful for the measurements we wish to make, particularly when written in the form

$$\frac{d\ln P}{d(1/T)} = \frac{L}{R} \tag{16}$$

If L is constant over the range of pressures and temperatures used, then (16) is readily integrated to give,

$$\ln P = -\frac{L_{\text{av}}}{R}\frac{1}{T} + \text{constant} \tag{17}$$

Compare this expression with the actual relationship between log P and $1/T$ given in Fig. 3 for water. To investigate how L varies with T we may write (16) in incremental form

$$\Delta \ln P = -\frac{L}{R}\Delta(1/T) \tag{18}$$

and determine the average L over small specific ranges of temperature (T_1 and T_2) and pressure (P_1 and P_2), so that

$$\ln P_2 - \ln P_1 = \frac{-L(T_2, T_1)}{R}\left(\frac{1}{T_2}\right) - \left(\frac{1}{T_1}\right) \tag{19}$$

(see Roberts and Miller[2] for further development of this discussion).

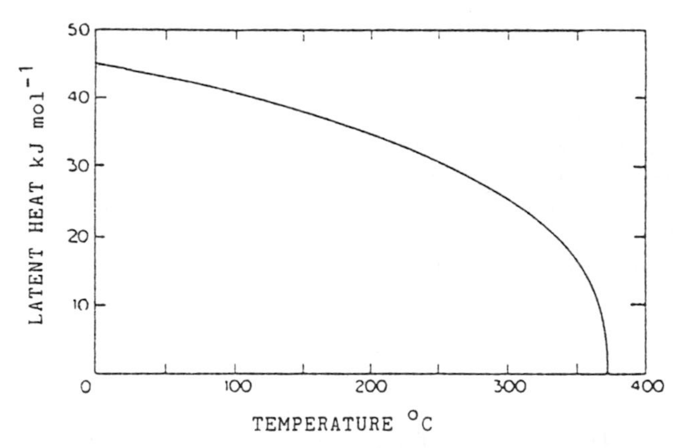

FIG.2 Graph showing A nonlinear relationship between the latent heat of vaporisation of water and temperature

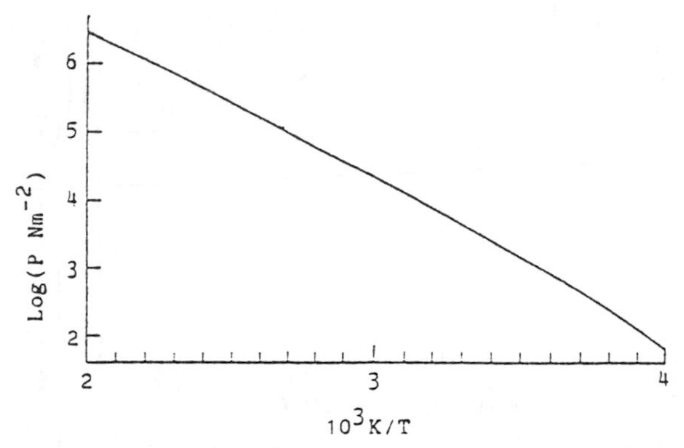

FIG.3 Graph of $\ln P$ versus $1/T$ between 250 K and 500 K (-23°C and 227°C) for water

Experimental investigation

The total latent heat of vaporisation is usally measured at the vapor pressure of the substance concerned. A simple way of doing this is to arbitrarily fix the external pressure and then to raise the temperature of the liquid until it begins to boil. At this point the vapor pressure of the liquid equals the externally applied pressure. Then L may be determined directly as the total heat supplied by calculating the amount of electrical energy during boiling required to vaporise one mole of the liquid – a well known and widely used method. Alternatively we may use a set-up similar to the one described by the diagram shown in Fig. 4 and study how vapor pressure varies with temperature by varying the boiling point through the use of a vacuum pump to reduce pressure. This will not only enable us to verify the expressions relating L, T, and P, but also use them to calculate L.

Software has been written and the computer used to make this interaction of theory and experiment as effective as possible. Two solid state transducers have been incorporated into the experimental setup as shown to measure pressure and temperature in parallel with the mercury manometer and mercury thermometer. The pressure transducer used was the modestly priced LX0603 with a linear operating range of $0-2$ atm. We used the Griffin unit based on this sensor for reasons of convenience and availability, and the thermistor GM472 with an operating range from -80 to 125°C.

The output from the pressure transducer could be fed straight into the AD converter of a BBC microcomputer. That from the

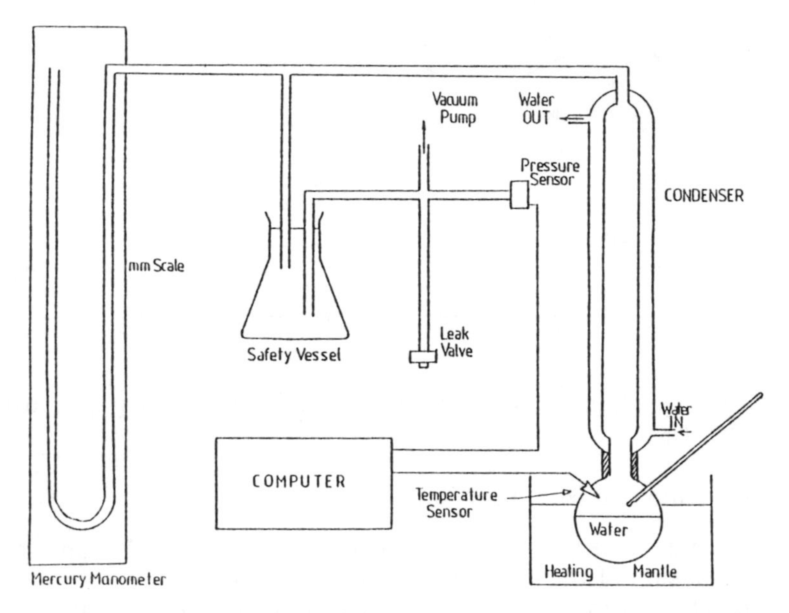

FIG.4 Diagram of the experimental setup used to study vapor pressure and the latent heat of vaporisation

thermistor was fed to the AD converter via a simple potential divider arrangement. Figure 5 gives a block diagram of the computer and interface and Fig. 6 the flow diagram for program VAPOUR. This setup enabled us to work in the temperature range between 50 and 100°C.

First, the software leads the student through a series of check procedures and then through the steps to reduce pressure and bring the liquid gently to the boil. After this the student is instructed to type in values of P and T from the mercury manometer and thermometer respectively whilst the computer automatically records the corresponding voltages from the transducers. The barometric pressure in the room is also requested and typed in. The procedure is repeated for several values of T chosen by the student between about 50°C and 100°C and the data stored in a student file. On completion, the data are presented to the student and various methods of analysis suggested by way of interpreting the data:

(i) Do you want a graph of P against T?

(ii) Do you want a graph of $\ln P$ against $1/T$? In this case

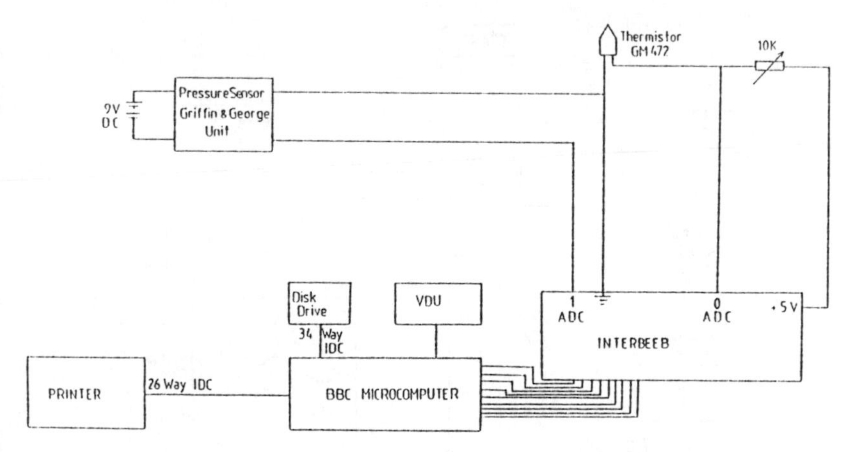

FIG.5 Block diagram of computer and interface

comparison is made with equation (17) and L_{av} calculated from the slope.

(iii) Do you want L calculated between different values of temperature? In this case the student chooses pairs of results of pressure and temperature and uses equation (19) to calculate L at $T = \frac{1}{2}(T_1 + T_2)$. This enables the student to appreciate how latent heat varies with temperature according to equation (19).

Finally, a data file containing standard values of T and P for water from Kaye and Laby[3] is presented to the student who may then repeat procedures (i) to (iii) on this data and compare the results with his experimental values.

In conclusion, the properties and uses of the transducers are investigated. The computer first plots calibration curves of T against the voltage (V_t) from the thermistor and P against the voltage from the pressure transducer. The nonlinearity of the first calibration curve is rectified by plotting T against $1/\ln V_t$.

The student is now invited to choose any arbitrary temperature and bring the water to the boil as before. Having calibrated its transducers the computer can now continuously monitor P and T in the reaction vessel so that the student can visualise more impressively how these parameters vary during heating and boiling and gain a richer understanding of the process of vaporisation.

Typical experimental results may be obtained from the authors.

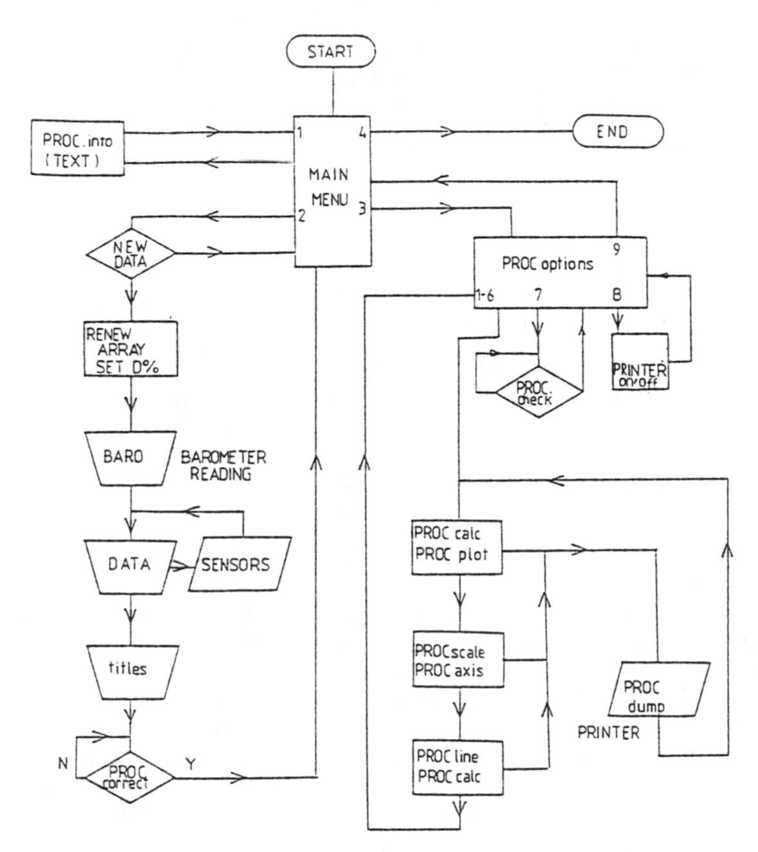

FIG.6 Flow diagram for program VAPOUR for the BBC microcomputer

References

1 S. Glasstone, *Textbook of Physical Chemistry*, Macmillan (1956), 2nd Edition.
2 J.K. Roberts and A.R. Miller, *Introduction to Heat and Thermodynamics*, Blackie (1951), 4th Edition.
3 G.W.C. Kaye and T.H. Laby, *Tables of Physical and Chemical Constants*, Longman (1973), 14th Edition.

Temperature control of a metal by software methods

A.R. Barnett
Department of Physics, University of Manchester, Manchester, M13 9PL, UK

Abstract: A Zenith 16-bit computer has been used as the decision element in a feedback method for achieving a steady temperature. Earlier work[1] made use of a BBC microcomputer. The experiment is part of a second-year course at Manchester to introduce both this style of computing and the general methods of feedback control.

Introduction

Morgan, Rosell, and McClean[1] in the first volume of this series, described a temperature stability experiment in which a BBC microcomputer program controled a heater through the user-port interface. Excellent screen photographs illustrate that work and the principles of feedback control. In our work at Manchester we decided to use a 16-bit machine and chose the Zenith AT 241. This decision also educated the staff in the difficulties of such an attempt and a summary of these will be given.

The object of the experiment is to stabilise the temperature of a block of metal heated by a 20-W resistor fed from a variable constant current source. The thermometer was an AD590 sensor (RS308-809) whose current output varies at the rate of 1 μA per °K. The feedback loop includes the computer running a FORTRAN or BASIC program and communication is achieved through an ADC/DAC card. A block diagram is given in Fig. 1. The student's task is to complete the details of a subroutine that makes decisions on the current to be fed to the heater. They are required to produce a write-up that should include graphs of the response of the system to various feedback parameters, comments on good values for the four control variables, and some information on the best stability obtained (about 0.2 deg).

A related experiment used a solid-state (Peltier) heat pump and a thermal reservoir. With this, and an additional current-controlled current source, it proved easy to drive the temperature of a thin sheet of metal down to 3° or 4° K and to stabilise it there.

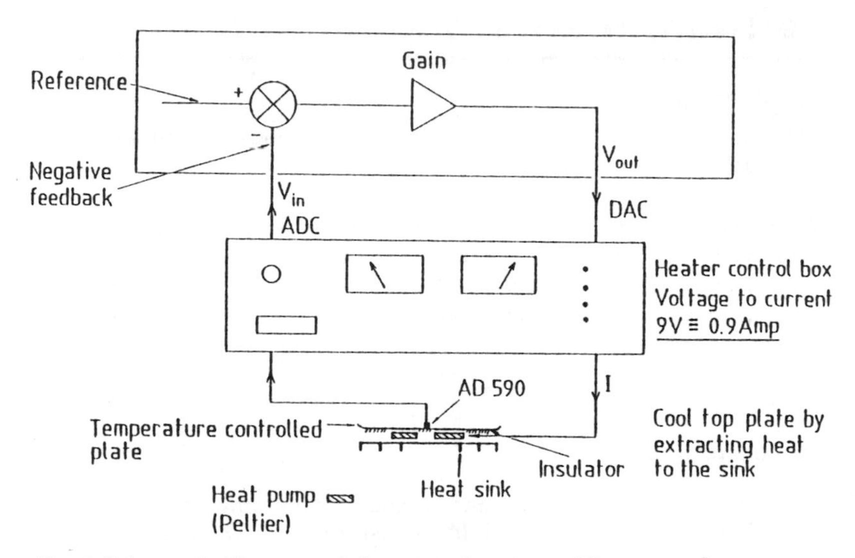

FIG.1 Schematic diagram of the control system. The upper box represents the computer program, interfaced through a DAC and an ADC to the heater control box

Eventually the reservoir temperature rose; it was reduced by stabilising at 50° or 60°K for a few minutes. Future improvements include adding thermal mass to the heat sink, such as a water-filled chamber.

Equipment

The Zeniths were fitted with an ADDA card that had one DAC (accepting 12 bits, 0-4095 and supplying $0 - 9\,V$) and 15 ADC inputs (also 12 bits and $0 - 9\,V$). In practice the noise level affected the last bit or two and the inputs were read usually 20 times and averaged. The method used appears at the end of the listing (Fig. 2, lines 1780-2060). The FORTRAN used was Lahey F77L[2] which has proved an excellent choice. For student use Lahey now offer Personal FORTRAN at $95 (and educational discounts) and I recommend it without reservation. The next problem is graphics and we ended up with the VDI drivers (IBM) and wrote our own. There is every reason to avoid the abyssmal CGA graphics offered on these PCs – at 200 lines their resolution is worse than the BBCs. We chose EGA graphics (350 lines) and linked the F77L to it. Nevertheless the BASIC used only supported the CGA graphics. It was supplied by M-TEC[3] and is in other respects an excellent product. (M-TEC released an EGA version in 1988). The machines were fitted with a switching graphics card; it supplies CGA or EGA screens to the same monitor and in these days of proliferating 'standards' such a card is essential.

```
 10 REM PC276 24/11/86 - 7/3/87 (ZENITH)
 20 REM "HOTEXPT" 19.3.87  TO BE MODIFIED TO CONTROL
 30 REM Temperature Control of METAL BLOCK
 40 REM Bham Exhibition 17,18/9/87
 50
 60 REM Version modified to cool as well using HEAT PUMP
 70 :
 80 ON ERROR PROCdaout(0.05):END
 90 :
100 REM R.Morgan + J.Rosell + W.McClean          &    ARB
110 REM Physics Experiments and Projects for Students
120 REM C.Isenberg and S.Chomet, Eds. Newman 1985, pp127-153
130 :
140 OSCLI"FX16,2"
150 *KEY4 OSCLI"FX16,2":L%=1:PROCcalib|M
160 *KEY1 55.0 |M 3 |M 0.05 |M 20 |M 1.0 |M
170 *KEY2 50.0 |M 3 |M 0.02 |M 10 |M 2.0 |M
180 *KEY3 14.0 |M 3 |M 0.02 |M  3 |M 2.0 |M
190 C%=6:D%=7
200 REM 2 adval channel numbers
210 REM adval LABELLED 0 1 2 3 ... !!!!
220 @% = &20204
230 MODE 3
240 olddiff=0
250 Iswitch=0
260 Z%=0
270 Vint  =0
280 Vref  =9.00
290 sample=100
300 REM nominal sample time IN centisec
310 average=20
320 step=4
330 Gain=1.0
340 TrueT%=FALSE
350 roomT =20              :V40=6.286 :V20=5.912
360 Zen=780/1024           :REM    Conversion of BBC Y scale to Zenith
370 Tmean =50
380 Trange=20
390 Tstp  =5
400 Xpos=60:Ypos=13
410 :
420 PROCTemp
430 PRINT"Temperature now is ";T
440 PRINT"REQUIRED Block Temp (C)"
450 INPUT reqT
460 COOL%=(reqT<20)
470 IF COOL% THEN cool=-1 ELSE cool=1
480 IF COOL% THEN Tmean=12:Trange=10:Tstp=1
490 PRINT'"REQUIRED Bandwith (deg C)"
500 PRINT'"Heater full on IF T is more than C deg lower than required temp"
510 PRINT"If too small (e.g. 0.1) => Bang-Bang"
520 INPUT band
530 Kprop=Vref/band
540 PRINT'"REQUIRED Integral const (0-3)"
550 PRINT"If too large get T oscillations"
560 INPUT Kint
570 PRINT'"REQUIRED Differential const (10-40)"
580 PRINT"Reduces the violence of the system"'"responses to control signals"
590 INPUT Kderiv
600 INPUT "Enter Gain of System 0.5 - 5.0  " Gain
```

```
 610 IF Gain < 0.1 THEN Gain=0.1
 620 :
 630 VDU23,1,0;0;0;0;
 640      REM   GO STRAIGHT INTO CONTROL STRATEGY
 650 MODE 0
 660 VDU19,1,0,0,0,0
 670 REM needed to cope with the EGA system
 680 VDU 29,0;400;
 690 PROCaxis
 700 REPEAT
 710   PROCTemp
 720   MOVE 120,(T-Tmean)*scale
 730   FOR X%=120 TO 1279 STEP step
 740     REM add Temp change
 750     key=INKEY(0)
 760     IF key=138 THEN Trange=20
 770     IF key=139 THEN Trange=60
 780     TIME=0
 790     PROCTemp
 800     PROCcalc
 810     PROCcontrol
 820     DRAW X%,(T-Tmean)*scale
 830     PRINTTAB(Xpos-20,Ypos+11) TIME;
 840     REPEAT
 850       PROCTemp
 860       PROCcalc
 870       PROCcontrol
 880     UNTIL TIME>=sample
 890   NEXT
 900 UNTIL FALSE
 910 END
 920 :
 930 REM********************************
 940 DEF PROCTemp
 950 nowV=0
 960 FOR I%=1 TO average
 970   PROCadval(C%)
 980   nowV=nowV + AD
 990 NEXT
1000 nowV=nowV/average
1010 @%=&20304
1020 PRINT TAB(Xpos,Ypos+2)nowV;"  NEW V"
1030 T=roomT*(nowV-V20)/(V40-V20)+20
1040 @%=&20204
1050 PRINT TAB(Xpos,Ypos+3)T;"  NEW TEMP"
1060 PRINTTAB(Xpos,Ypos-2)"Gain     ";Gain
1070 ENDPROC
1080 :
1090 DEF PROCcalc
1100 REM++++++++++++++++++++++++++++++++++++++++
1110 REM this is where you decide on the control
1120 REM++++++++++++++++++++++++++++++++++++++++
1130 VDU23,1,0;0;0;0;
1140 diff        = reqT-T
1150 Vprop = Kprop*diff
1160 Vdiff = Kderiv*(diff-olddiff)
1170 Vint = Kint*diff+Vint
1180 IF ABS(diff)<0.25 THEN Z%=Z%+1
1190 IF Z% =>2 THEN Iswitch=1
1200 Vint=Vint*Iswitch
1210 olddiff = diff
1220 VOUT = Vprop+Vint+Vdiff
1230 VOUT = VOUT*cool
1240 :
1250 PRINTTAB(Xpos,Ypos+8)"VOUT      ";VOUT
1260 PRINTTAB(Xpos,Ypos+5)"Vprop     ";Vprop
1270 PRINTTAB(Xpos,Ypos+6)"Vint      ";Vint
1280 PRINTTAB(Xpos,Ypos+7)"Vdiff     ";Vdiff
1290 ENDPROC
1300 :
1310 DEF PROCcontrol
1320 IF VOUT > 9.00 THEN VOUT=9.00
1330 IF VOUT <=0.05 THEN VOUT=0.05
1340 PROCdaout(VOUT)
1350 ENDPROC
1360 :
1370 DEF PROCaxis
1380 REM Tmean=reqT:Trange=dT:stp=1
1390 REM Tmean=50:Trange=60:Tstp=5  typically
1400 scale=1024*Zen/Trange
1410 Tmean=reqT
1420 REM ARB trial
1430 VDU5
```

```
1440 @%=2
1450 Tr2=Trange/2
1460 FOR T% = Tmean-Tr2 TO Tmean+Tr2 STEP Tstp
1470   M%=(T%-Tmean)*scale
1480   MOVE 60,M%:DRAW80,M%:MOVE 0,M%-16
1490   PRINT T%
1500 NEXT
1510 T%=(reqT-Tmean)*scale
1520 MOVE 120,T%:PLOT5+16,1280,T%
1530 REM desired temp
1540 MOVE 70,500*Zen:PRINT"deg"
1550 MOVE 70,460*Zen:PRINT" C"
1560 MOVE 60,-400:DRAW 60,400
1570 MOVE 60,0:DRAW 1279,0:DRAW 1279,50
1580 timeX=sample*1160/100/step
1590 MOVE 1100,-32:PRINT;timeX;" sec"
1600 FOR T%=0 TO timeX STEP 10
1610   X%=T%*1160/timeX
1620   MOVE 120+X%,10:DRAW 120+X%,-10
1630   REM IF X% MOD 100 THEN MOVE 120+X%,20:DRAW 120+X%,-20
1640 NEXT
1650 xpos=900
1660 @%=&20204
1670 MOVE xpos,380:PRINT"required T = ";reqT
1680 MOVE xpos,320:PRINT"prop + int + der"
1690 MOVE xpos,280:PRINT"bandwidth ";band;" deg C"
1700 MOVE xpos,240:PRINT"integ con ";Kint
1710 MOVE xpos,200:PRINT"deriv con ";Kderiv
1720 REM MOVE 40,-400
1730 VDU4
1740 ENDPROC
1750 :
1760 REM end of term Xmas 1986
1770 :
1780 REM ****************PSEUDO ADVAL PROCEDURE************
1790 DEF PROCadval(CHANNEL)
1800 ADPORT=&278
1810 Convert=9.00/&FFF
1820 PUT ADPORT,CHANNEL
1830 TEMP=GET(ADPORT+3)
1840 FOR JCOUNT=0 TO 6
1850   TEMP=GET(ADPORT+4)
1860 NEXT
1870 FOR JCOUNT=0 TO 6
1880   TEMP=GET(ADPORT+5)
1890 NEXT
1900 TEMP=GET(ADPORT+2)
1910 HBYTE=TEMP MOD 16
1920 LBYTE=GET(ADPORT+1)
1930 AD=HBYTE*&100+LBYTE
1940 IF L%=0 THEN PRINT TAB(Xpos,Ypos+10)AD,TAB(Xpos+10,Ypos+10)"AD";
1950 AD=AD*Convert
1960 ENDPROC
1970 :
1980 REM *************OUTPUT TO D/A CONVERTER***************
1990 DEF PROCdaout(VoltageOutput)
2000 DAPORT=&27E
2010 @%=&20204
2020 Convert=&FFF/9.00
2030 VO%=VoltageOutput*Convert
2040  PUT DAPORT,(VO% MOD &100)
2050  PUT (DAPORT+1),(VO% DIV &100)
2060 ENDPROC
2070 :
2080 DEF PROCcalib
2090 VDU23,1,0;0;0;0;
2100 REPEAT:PROCadval(C%):@%=&20306:PRINT"Sensor "C%;" = ";AD;" ";
2110   @%=10:T%=AD*&FFF/9.00:PRINTT%:Z=INKEY(100)
2120 UNTIL FALSE
2130 REM A.R.BARNETT , Manchester March 1987
```

FIG.2 BBC BASIC program for the computer control. It runs on the Zenith 16-bit PC machine

The circuit of the electronics consisted of a voltage-controlled constant-current source delivering $0 - 1\,\mathrm{A}$ into $2\,\Omega$. The voltage range is 0 – 9 V, supplied by the DAC. The meter circuit, Fig. 3, measures the voltage developed across the 20 $k\Omega$ resistor. It is calibrated with a standard current source and serves as the temperature reference over the range 0-100°C.

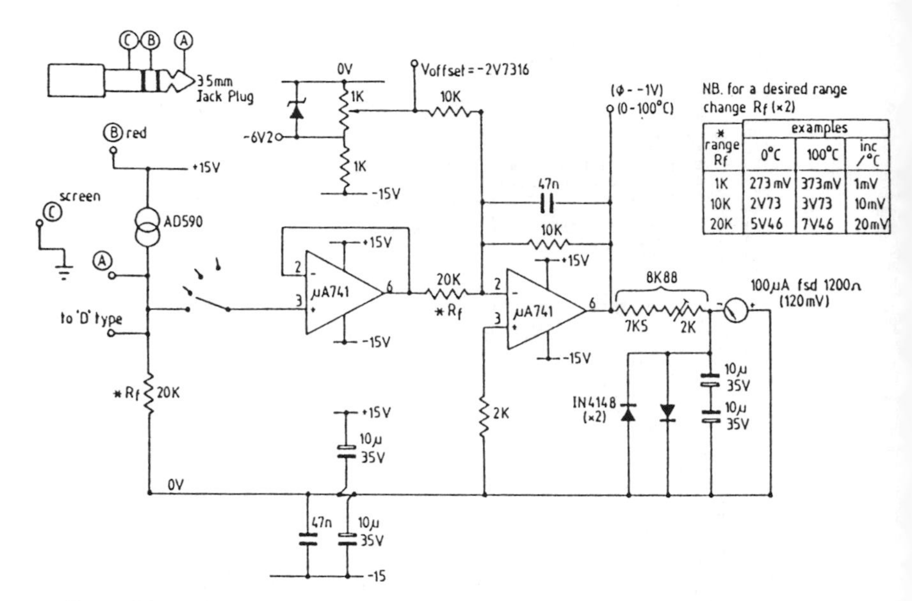

FIG.3 Meter circuit that acts as a temperature reference (by precisely setting $V_{\rm offset}$). The AD590 supplies $1\mu \mathrm{AK}^{-1}$ and the circuit, switchable to one of 15 inputs, provides a direct-reading of temperature

Student approach

The students were supplied with a program that lacked only the control algorithm (lines 1140-1230 of Fig. 2). They had to devise a control strategy that included proportional control, $V_{\rm prop}$, integral control, $V_{\rm int}$, differential control, $V_{\rm diff}$, and some gain, G. The suggested form of the total control voltage was thus:

$$V_{\rm tot} = G(V_{\rm prop} + V_{\rm int} + V_{\rm diff})$$

V_{prop} is approximated by $K_1\Delta T$, where $\Delta T = T_{reqd} - T$; its function is to reduce the heating (or cooling) when ΔT is small enough (typically 3°) and the required temperature is being approached. V_{int} is approximated by

$$K_2 \int (\Delta T/dt)dt = K_2 \sum \Delta T$$

This is achieved by setting $V_{int} = 0$ and then at each time step $V_{int} = V_{int} + K_2\Delta T$. The inclusion of V_{int} helps to compensate for heat losses/gains to the environment by providing some power even when $\Delta T = 0$. V_{diff} is approximated by $K_3\Delta T/dt$ and the intention is to prevent sudden excursions being over dominant by opposing them. It was found that various subtleties to do with averaging the differential control were advisible as noise tended to intrude on these values. The program displayed current values on screen and an empirical adjustment was possible when, for example, the gain or K_2 were too large and the control system oscillated. A much more sophisticated approach is to try to relate the values of K_1, K_2, K_3 and G to the thermal time-constants of the physical system (which can be readily measured). An accessible discussion of the principles involved, including the relationship to Nyquist Diagrams, is to be found in Forgan.[4] The last step is to predict the variation of temperature with time, knowing the frequency response of the components (thermal time-constants), by using Fourier transform techniques. See also the Engineer's approach.[5]

Comments

The results that were obtained were in agreement with published results.[1] The system used had a very different ratio of thermal capacities of the heater and the heated block (oven) and stabilisation in our system was a more delicate matter; it was easier to get it wrong! Forgan's recommendations[4] have been bult into the next generation of equipment. It is useful to build in additional thermal delays, e.g., immersing the AD590 thermometer in oil, to introduce extra phase shifts in the stabilisation characteristics so that more complicated strategies are required for stability. Another feature investigated by some students was the effect of having the control signal proportional to the output current, I, instead of taking the square root so that the proportionality is to the power. The introduction of nonlinearity in this way did not worsen the degree of control and the best results obtained (±0.1°C over 40 min) in fact used it.

The very different behaviour of the heater-block combination and the Peltier heat pump is worth noting. The heat pump drives heat in or out at a similar rate (dependent on the current), whereas the heater can drive the temeprature up rapidly, it only cools slowly by heat exchange with the environment (convection, radiation, and a little conduction). Thus it is worth considering if different K_3 values are appropriate when ΔT is positive and when it is negative. Our present conclusions are unclear.

A design improvement to the meter circuit of Fig. 3 would be to offset the AD590 current and increase the input gain so as to present a larger voltage swing for 0-100°C to the ADC. The 2 V change from 5.5 V to 7.5 V would become a 8 V change and the noise problems in the ADC would be less apparent. The stability of the offset current would then become important instead.

In no way does the experiment OR *require* a 16 bit machine and in truth the setting-up on a BBC micro is a much easier task because the interfacing is already done. However we regard the exercise as very worthwhile: the Zenith can be programmed in FORTRAN, PASCAL, or C as well as BASIC as required, and the introduction to 'computing as it will be in the outside world' was regarded positively by a number of students. The concept of software control was strongly reinforced and none of the students doubts the power of the methods demonstrated in the experiment.

Acknowledgments

The realisation of the project described here and the creation of the entire Zenith Laboratory would have been impossible but for the willing and dedicated participation of Jim Allen, Head of the Electronics Workshop in the Physics Department. I am most grateful for his efforts. I thank also Ian Morison, Lecturer in Physics, for his valuable assitance during the course.

References

1 R. Morgan, J. Rosell, and W. McClean, *Physics Experiments and Projects for Students*, C. Isenberg and S. Chomet eds., Newman-Hemisphere, pp127-153 (1985)
2 Lahey Computer Systems Inc., P.O. Box 6091, Incline Village, Nevada 89450, USA
3 M-TEC Computer Services (UK), Olland Road, Reepham, Norfolk NR10 4FL
4 E.T. Forgan, Cryogenics **10** (1975)
5 S.A. Marshall, *Introduction to Control Theory*, Macmillan (19780

A large-scale thermocouple holds up several kilograms!*

S.M. Kay
Royal Holloway and Bedford New College, University of London, Egham, Surrey, UK

Introduction

When the junctions of a large scale thermocouple are maintained at different temperatures the current generated produces a magnetic field sufficient to magnetise two mild steel blocks. Magnetic forces up to 100 N can readily be generated in this way. This effect is used as the basis of an investigation carried out by first year undergraduates.

Experimental details

The thermocouple consists of a bent copper rod, 10 mm in diameter, bridged by a short bar of constantan (60 copper, 40 nickel; see Fig. 1). The copper loop is approximately 175 × 20 mm. One end of the copper is heated with a bunsen burner and the other end is kept cold with ice. Two milled-steel blocks, approximately 100 mm by 50 mm and 22 mm deep encase the copper rod. The blocks are insulated from the copper (brown paper suffices), and the steel surfaces that are in contact with each other (see Fig. 1c) are milled flat and polished so that they are in as close contact as possible. It is important that the air gap between the blocks is a minimum (see the discussion below).

Initially, the steel blocks are held together with a screw but once a sufficient temperature difference is established the blocks become magnetised and the screw can be released. (For safety, wire can be wound loosely round the blocks to support the lower block when they separate.)

Measurements can now be made of the magnetic force produced, as a function of the temperature difference across the junctions,

* **Reprinted with permission from the European Journal of Physics**

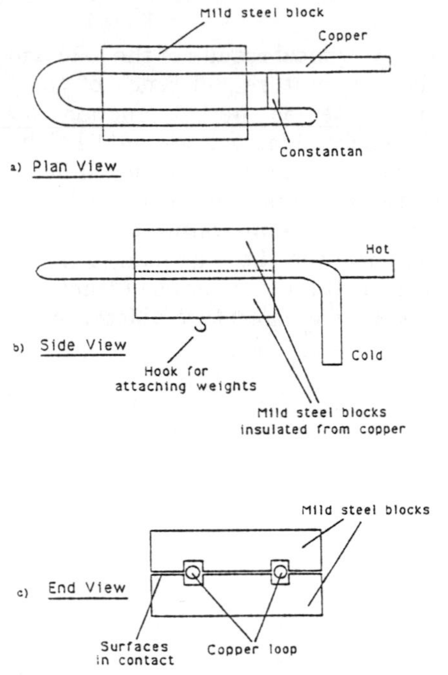

FIG.1 Thermocouple magnetism apparatus

by hanging weights from the lower block. The temperatures of the junctions are measured using small scale thermocouples fitted into small holes drilled into the copper rod where it joins the constantan.

Students soon discover that the force developed depends on whether measurements are taken as the hot junction is heated up or as it cools down, i.e., that there is hysteresis in the magnetisation. They also discover that the force depends on how well the blocks fit together. These observations lead to consideration of questions of reproducibility and how best to make systematic measurements. The best way we found to obtain reproducible results was to start with the 'hot' junction at room temperature and to heat up the junction each time to the same temperature (100°C was found sufficient for the blocks to be well magnetised). A known load was then added and the temperatures of the two junctions recorded at the moment of separation of the blocks under this load.

Results

Figure 2a shows some typical results of the load added to the lower block against the temperature difference of the junctions when separation occurred taken as the hot junction cools. The largest load that could be safely added was around 9 kg! It is seen that the plot extrapolates at near zero temperature difference to a load of 2.5 kg. By comparison, it was found that with the blocks initially unmagnetised a minimum temperature difference of around 35°C was needed to support even the 1-kg load of the lower block.

Figure 2b shows the total force applied (including the weight of the lower block) against temperature difference.

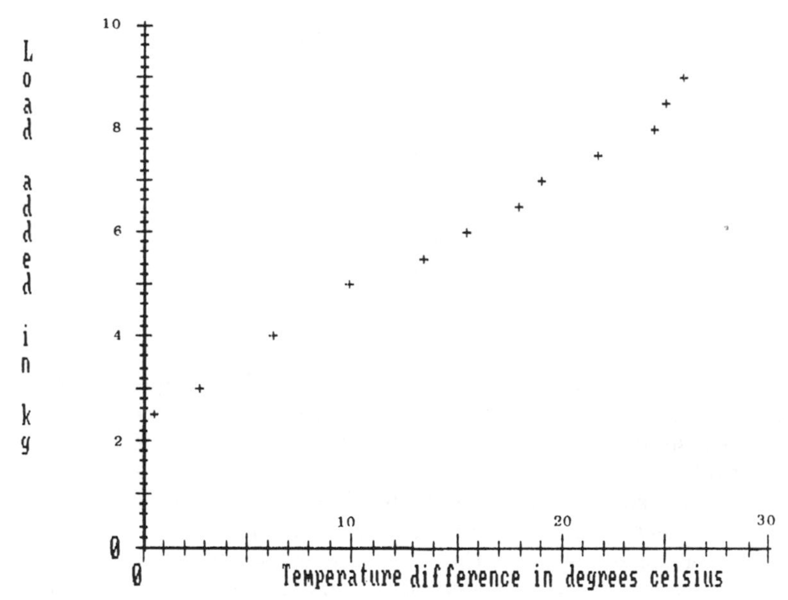

FIG.2a Experimental data of the added load against temperature difference

Theoretical discussion

The magnetic force produced can be calculated by considering the steel blocks as forming a magnetic circuit around a single current loop carrying the thermoelectrically produced current I (see Fig.

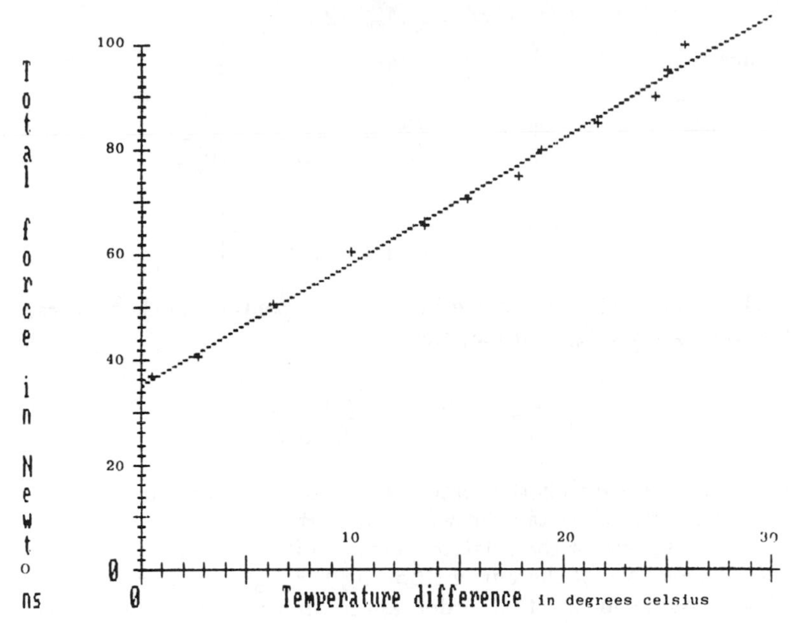

FIG.2b Experimental data of the total force against temperature difference

3). The magnetic induction in the air gap is given by

$$B_{\text{gap}} = I/A_{\text{gap}}R$$

where A_{gap} is the cross-sectional area of the gap and R the magnetic reluctance is given by

$$R = \left[\frac{\ell_{\text{steel}}}{\mu_{\text{steel}}\mu_0 A_{\text{steel}}} + \frac{\ell_{\text{gap}}}{\mu_0 A_{\text{gap}}}\right]$$

where ℓ_{steel} and ℓ_{gap} are the lengths in the steel and air gap, respectively, A_{steel} and A_{gap} are the corresponding cross-sectional areas which in this case are equal, and μ_{steel} is the relative magnetic permeability of steel (the theory of magnetic circuits is discussed,for example, by Lorrain and Corson[1]).

The magnetic force produced across the gap due to the magnetic flux is given by

$$F = B_{\text{gap}}^2 A_{\text{gap}} / 2\mu_0.$$

Hence

$$F = \frac{A_{\text{gap}}}{2\mu_0} \left[\frac{\mu_o A_{\text{gap}}}{(\ell_{\text{steel}}/\mu_{\text{steel}}) + \ell_{\text{gap}}} \right]^2 \frac{I^2}{A_{\text{gap}}^2}$$

$$= \frac{A_{\text{gap}} \mu_0 I^2}{2} \left[\frac{1}{\ell_{\text{steel}}/\mu_{\text{steel}} + \ell_{\text{gap}}} \right]^2$$

If the mild steel surfaces are in initimate contact so that we can assume $\ell_{\text{gap}} << \ell_{\text{steel}}/\mu_{\text{steel}}$, then

$$F = \frac{\mu_o A_{\text{gap}} I^2}{2} \left(\frac{\mu_{\text{steel}}}{\ell_{\text{steel}}} \right)^2$$

From this we see that the force generated is proportional to
(a) the area of contact between the surfaces,
(b) the square of the relative permeability of steel,
(c) the square of the current flowing through the loop.
If we assume in the first order approximation that the current flowing through the loop is proportional to the temperature difference ΔT across the junctions, then the force generated will be proportional to $(\Delta T)^2$.

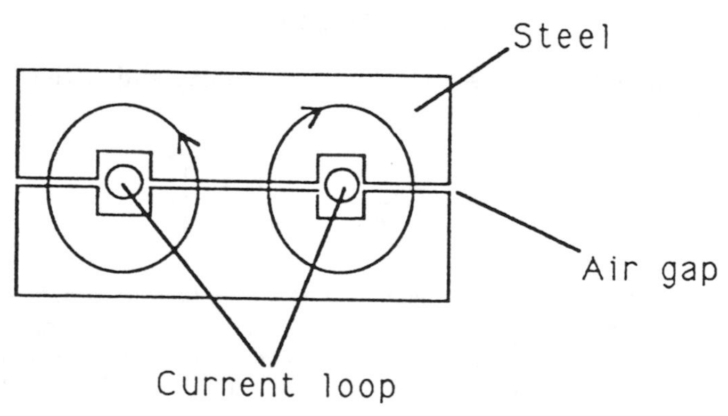

FIG.3 Magnetic circuit

Application to the experimental situation

For the blocks used in this experiment, ℓ_{steel} (average = 90 mm and A_{gap} = 1350 mm^2. The main imponderable is the relative permeability of the steel blocks. This quantity is very dependent on the material. It also depends of course on the strength of the magnetic field and on the direction of magnetisation. For mild steel, typical values are between 1000 and 3000. Taking this range for the value of μ_{steel} and including two current loops (see Fig. 3), the calculated value of B_{gap} lies between $0.01I$ and $0.03I$ T and F between $0.2I^2$ and $1.8I^2$ N. The maximum measured force of 100 N then corresponds to a current I of between 7 and 22 A.

There is no easy way of measuring the current. A clip-on ammeter used for measuring d.c. currents in wires would be ideal, but unfortunately they are not readily available to clip over as large a diameter as 10 mm.

Discussion

Referring to the experimental data in Fig. 2b we note that the measured force remained large even at very small temperature differences. In the theory given above it was assumed that μ_{steel} is constant. However, during the demagnetisation part of the hysteresis cycle μ increases as B decreases towards zero. This will have the effect of increasing the force generated at small ΔT values.

We also note that over the range of temperature differences investigated the force varied fairly linearly with ΔT, whereas the theory predicts a $(\Delta T)^2$ dependence. However, in the theory various factors have been neglected. We note that the current in the loop is given by the thermoelectrically generated EMF divided by the total resistance of the loop, and in the derivation we neglected:

(i) any nonlinearity in the variation of the EMF with temperature difference; this, however, will be small for a copper-constantan thermocouple over the temperature differences produced here

(ii) the variation in the resistance of copper and constantan due to changes of resistivity with temperature; over the temperature range covered, the increase is approximately 15%, which results in a corresponding decrease in the current, and a larger reduction in the force generated.

Probably more significantly, the length of the air gap has been neglected in the theory, i.e., ℓ_{gap} has been neglected in comparison with ℓ_{steel}/μ_{steel} in the expression for the magnetic reluctance R. Using the value of ℓ_{steel} of 90 mm and taking μ_{steel} to be 1000, this requires ℓ_{gap} to be much less than 0.1 mm. If ℓ_{gap} is equal to 0.1 mm the two terms in the expression for the magnetic reluctance will be comparable, so that R will increase by a factor of 2 and the force will decrease by a factor of 4. This explains why it is so important for the steel surfaces to be flat and polished.

Further investigations

The magnetic field generated in air by the current loop is small, but it can just be detected using a 200 turn search coil of area $2 \times 10^{-4}\,\mathrm{m}^2$. A 100°C temperature difference generates a field of approximately 1 mT (50 times the earth's magnetic field). However the presence of the magnetic field can be more readily demonstrated using a small compass needle. The copper-constantan loop, with the compass needle mounted inside the loop, is supported vertically (see Fig. 4), and the plane of the loop is aligned parallel to the earth's magnetic field. When a temperature difference is generated across the junctions, the compass needle is seen to rotate and to align itself with the direction of the thermoelectrically induced field perpendicular to the plane of the loop.

These observations are consistent with the magnitude of the currents estimated above. The field due to a rectangular loop of width 2a and length 2b carrying a current I is given by

$$B = \mu_0 I(a^2 + b^2)^{1/2}/\pi ab$$

. From the dimensions of the copper-constantan loop, $B = 4 \times 10^{-5} I\,\mathrm{T}$. Thus 1 mT would correspond to a current of 25 A.

Practical application of the effect

This effect is used as the basis of a flame failure device in gas installations. The pilot light heats one end of a thermocouple junction. The magnetic field produced energises an electromagnet that operates to keep the main gas inlet valve open. If the gas supply fails or the pilot light goes out, the thermocouple cools and the magnet is no longer energised. The gas valve closes and the

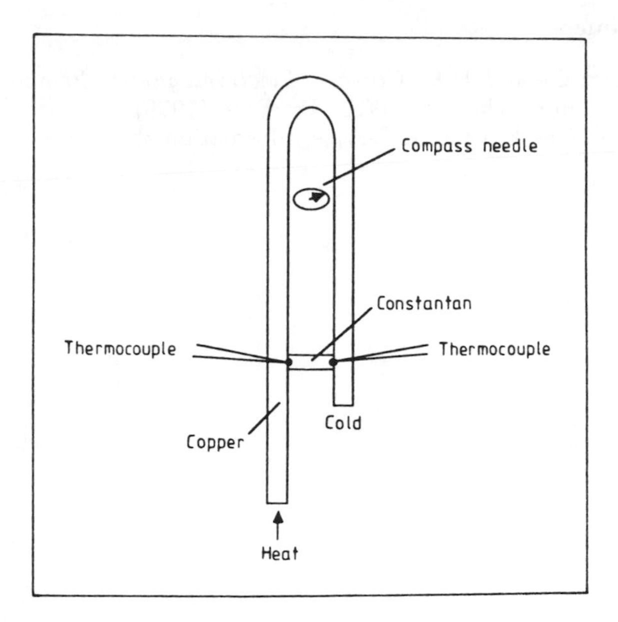

FIG.4 Arrangement for detecting the magnetic field

gas is automatically cut off to the main burner and to the pilot light.[2]

Conclusions

Students find this an interesting topic for investigation in which they learn about thermoelectric effects and magnetic circuit analysis, topics which are not necessarily covered in lectures. In the course of their investigation they are also made aware of the effects of hysteresis and are led to consider questions of reproducibility in measurement. Finally, they learn that basic physics principles find application in a commercial device.

References

1 P. Lorrain and D.R. Corson, *Electromagnets: Principles and Applications*, chap. 15, W.H. Freeman (1979)
2 British Gas Education Service information sheet

Determination of the Curie point of nickel by a resistance technique

R.A.L. SULLIVAN, A. DUNK, P.J. FORD,* AND R.N. HAMPTON
School of Physics, University of Bath, Claverton Down, Bath, BA2 7AY, UK

Abstract: This article describes an undergraduate laboratory experiment which investigates the transition from ferromagnetism to paramagnetism in a nominally pure nickel sample by means of an electrical-resistance technique. The temperature dependence of the resistance is measured in the neighborhood of the Curie temperature which is then determined by an elegant analysis of the data. The experiment demonstrates the usefulness of the BBC microcomputer both for collecting and analysing the data.**

Introduction

Despite the fact that the lodestone has been known since antiquity,[4] magnetism is a subject which is still not completely understood, although enormous progress has been made in the last few years [23]. Its study normally takes place in undergraduate physics courses and is discussed in standard textbooks on solid state physics.[17,11,21,1] Students are frequently confused by the numerous different forms in which magnetism can exist.[13]

In this article we describe an undergraduate experiment in which students examine the transition from ferromagnetism to paramagnetism in a nickel sample and thereby determine its Curie temperature. The electrical resistance R of a nominally pure nickel sample is measured as a function of temperature T in a range of ± 30 K of the Curie point. The data are collected with the aid of a BBC microcomputer, which is also used in the analysis. The experiment therefore provides students with useful experience in interfacing their apparatus to a microcomputer and in the writing of programs to analyse their data. Undergraduate experiments along

---

* **Physics Department, University of the Witwatersrand, Johannesburg, South Africa**

** **Reprinted with permission from the European Journal of Physics**

these lines have already appeared in the literature.[14,9] However, in this article we emphasise the value of the BBC microcomputer both for collecting and analysing the data as well as suggesting that the experiment provides a useful introduction to the study of critical phenomena.

The study of transport properties of ferromagnetic systems near the critical point was a prominent research area during the 1960s and the behavior of the electrical resistivity was fully discussed in a review article by Kawatra and Budnick.[15] The resistivity ρ has a characteristic change of slope or 'knee' at the Curie temperature T_c. More striking is the behavior of $d\rho/dT$ which shows a divergent behavior at T_c. The reason for this is complicated but results from changes in both the long- and short-range contributions to the spin disorder part of the total resistivity.[15]

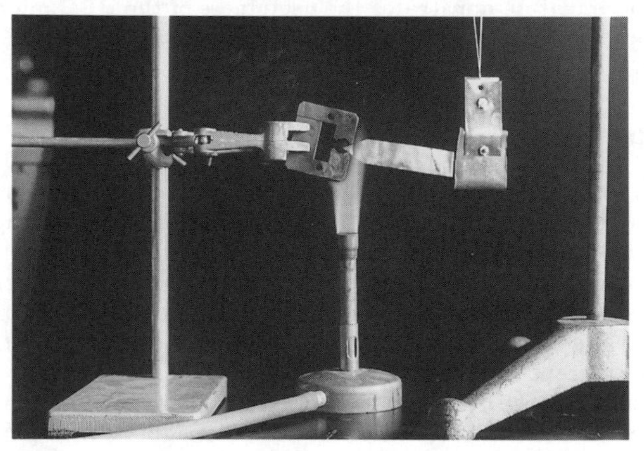

FIG.1 Simple demonstration of the Curie point of nickel

The three most common ferromagnetic materials are the third transition-metal elements iron, cobalt and nickel. Of these, nickel is the most convenient to study since it has a comparatively low Curie temperature (358°C) which can readily be reached with undergraduate laboratory equipment. The majority of students have never seen the transition from ferromagnetism occuring at the Curie temperature and to observe this they first carry out the short, simple demonstration experiment shown in Fig. 1.[3] Initially the nickel pointer is attracted to the permanent magnet as shown. Heat from the Bunsen burner is supplied to the nickel until it is above its Curie temperature. At this point it becomes paramagnetic and the force of attraction is greatly reduced. Hence the nickel falls out of the Bunsen flame and rapidly cools to below its Curie temperature. It thus regains its ferromagnetism and is attracted back to the permanent magnet and the cycle repeated. This continues for as long as the nickel is heated. The far end of the nickel pointer should be weighted with a heavy material, such as a piece of lead, so as to produce relatively undamped swings. We feel that the experiment could form the basis for an executive toy. There exist other 'toys' based on the Curie point of nickel.[3]

Apparatus

The apparatus used in the experiment is shown in Fig. 2(a) and (b). The furnace contains the nickel specimen which is in the form of a wire 0.25 mm in diameter having a total resistance at ambient temperature of about 8.8 Ω. The wire is wound on a hollow, grooved former, 27 mm outside diameter and 75 mm long. This is enclosed in a copper tube, 35 mm outside diameter and 430 mm long, to try to maintain uniformity of temperature. The temperature is measured with a chromel-alumel theromcouple in the centre of the former and the thermal e.m.f. converted to temperature. Further details of this conversion are discussed below. The furnace and its insulation are physically large, 30 × 30 × 47 cm, so that the time constant and the thermal capacity of the setup are also both large. The relative disposition of specimen, heater, and thermocouple are such that hysteresis is observed on rise and fall of temperature and is a cause of student comment and explanation. The actual heater in the furnace is under manual control only and the student can choose how fast to raise or lower the temperature. In fact a slow rise towards the Curie point allows considerable familiarisation and experimentation with the program *en route*.

The electronics unit used in the experiment is shown in Fig. 3. The temperrature of the specimen is measured by the thermocouple, the output of which is amplified approximately 60 times. A chopper-stabilised operational amplifier (7 650) is used so that the amplifier offset voltage and its attendant drift with temperature are entirely negligible. The circuit uses a differential configuration so that reasonable commonmode rejection is available to defeat the inevitable a.c. pickup and other signals so often prevalent in an undergraduate laboratory. The feedback resistor is bypassed (1 s time constant) to assist this process further.

In the event of accidental wrong connection, the output of the amplifier could approach either of the two rail voltages (±5 V) and hence a protection circuit is interposed between it and the analogue input to the BBC computer. This is produced by using one (bipolar) operational amplifier in the set of four provided by the 324

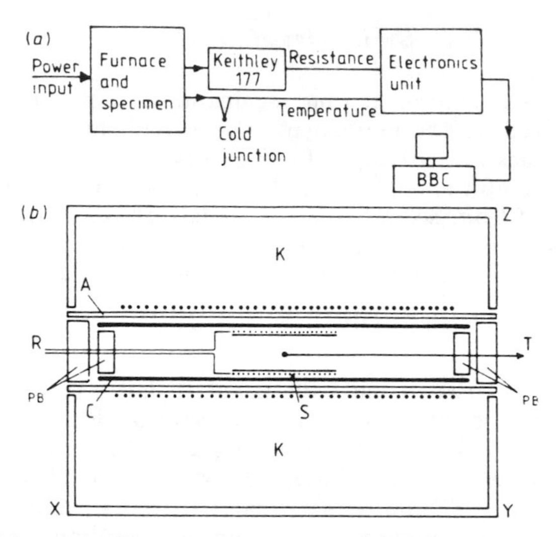

FIG.2 (a)Schematic diagram of apparatus. (b) Detailed view of the furnace and nickel specimen showing: copper tube C, specimen S wound around a grooved silica former and coated with heat resistant cement (Auto stick), Pyrophyllite bungs PB, Kaowool insulation K, alumina tube A around which are wound the heater coils, thermocouple T, and resistance leads R. The dimensions are XY = 47 cm, YZ = 30 cm.

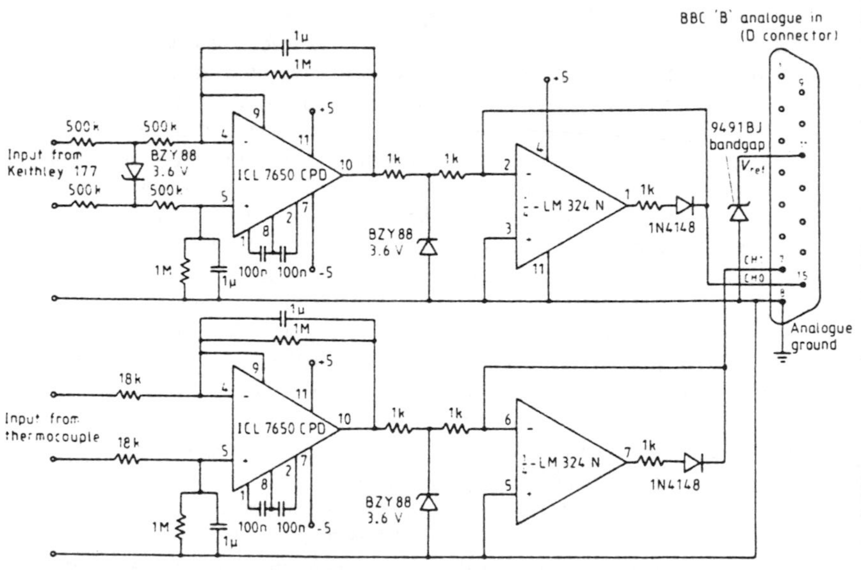

FIG.3 The electronics unit and interfacing arrangements used in this experiment

chip.[12] A further refinement to the analogue measurement facilities provided by the BBC computer is the addition of a bandgap reference diode wired in parallel with the existing reference terminals. This yields a much more accurate reference voltage of approximately +1.25 V with a temperature coefficient of less than 100 PPM K^{-1}.

The part of the circuit which deals with the resistance measurement follows on similar lines. A Keithley 177 digital multimeter (DMM) is used to measure the specimen resistance directly and a signal from the analogue output at the rear of the instrument is used to provide data to the electronics unit and hence to the computer. The analogue signal is negative going with -2 V output at full scale from a source impedance of $5\,\mathrm{k}\Omega$, so that it can be readily dealt with by another 7650 differential amplifier operated at unity gain. The analogue output of the Keithley can rise to ± 15 V if the instrument goes into over-range, the 7650 being protected against this eventuality by the Zener diode shown. In addition a protection circuit follows the 7650 before the signal is applied to the analogue-digital converter (ADC) of the computer.

Software for the BBC microcomputer

The ADC of the BBC (μ PD 7002) is used with the ADVAL command. Thus, the whole program can be written in BASIC, as speed is of little consequence in a system with long response time. The program is menu driven and allows the user various choices.

(i) To collect the resistance and temperature (R, T) data and to store them in RAM and also calculate during the collection dR/dT from the raw data. This calculation is performed by obtaining the best least-squares straight line through an odd number of (R, T) points and taking the calculated gradient of the midpoint as the gradient at the midpoint. This is done for all the (R, T) ponts except for the first and last few. Data collection may be carried out in a variety of ways, either by collecting a specified number of points or collecting over a specific temperature range or even for a certain time. In fact the speed of collection will depend on the number of readings averaged to obtain one (R, T) point. This number can be varied and the student is encouraged to use a second program (STADEV) to examine the statistics (mean, standard deviation of population and of mean and percentage error) of the incoming data and their dependence on the number of readings averaged. The limitations of the 10 to 12 bit accuracy can thus be minimised.

The conversion of the thermocouple voltage to temperature may be accomplished in ways of increasing sophistication. From the published thermocouple tables,[2,5] the approximate value of 40.87 μVdeg^{-1} and even 0.1°C near 360°C, is sufficiently accurate for our purposes. Polynomials of different complexity may also be used.[2,5,20] However, we have been unable to otbain any agreement with the thermocouple tables[2,5], using the formula given by Fox *et al.*,[9] which appears to us to be misprinted.

(ii) To smooth the (R, T) data by averaging over a window of $(2n + 1)$ points and allocating the average value to the midpoint of the $(2n + 1)$. This window width is then swept over all the points collected and new values for dR/dT calculated. The user can be encouraged to write alternative smoothing programs here and insert them , eg., by calculating the best straight lines (or even curves) through an odd number of points and allocating the calculated value of R_{mid} to the midpoint. It is also possible to do more than one cycle of smoothing. However care should be exercised that the window width is not so large that it smooths out the effect of the transition.

(iii) A plot or display choice may be made in which R against T and dR/dT against T are plotted on the screen for the range of T covered. Using a screen-dump these graphs may also be output onto paper.

(iv) The data can be printed out. This is not a much used facility as it is more convenient to process data within the BBC or to use a disk.

(v) The data can be stored on disk.

(vi) The data can be read fron disk for later processing.

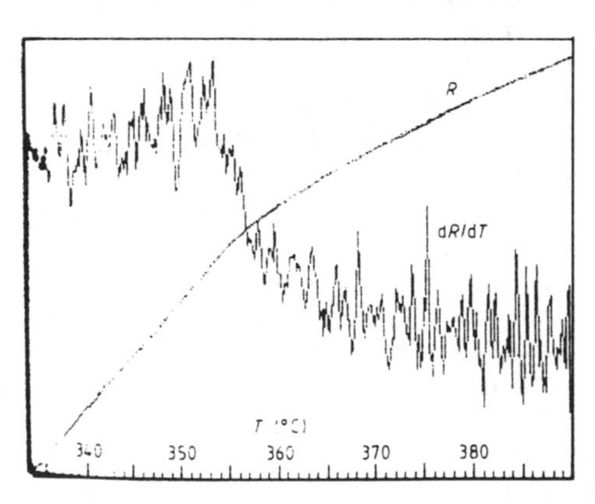

FIG.4 Typical results using 314 raw data points. R and dR/dT are on arbitrary scales

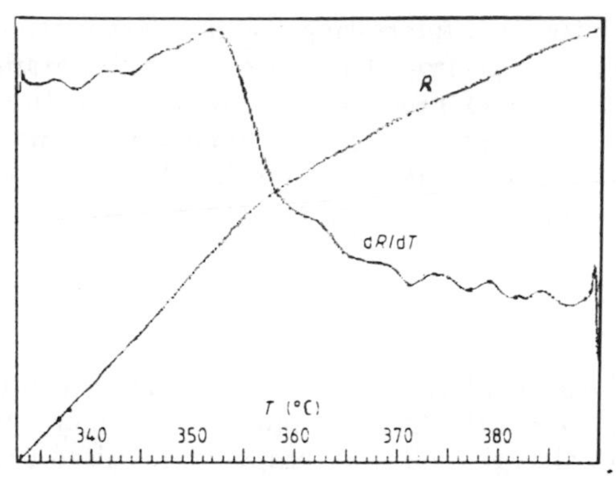

FIG.5 Typical results showing data from Fig. 4 smoothed four times using a nine-point smoothing routine

Results

Some typical results are shown in Figs. 4 and 5 from which it may be seen that smoothing is of great advantage. It is fairly clear that the form of the (R, T) plot just below the Curie point is a curve whose gradient steadily increases as T increases. Eventually a maximum is reached in dR/dT and thereafter it falls until it reaches a constant level. The question then arises: where is the Curie point? Here we should consider the very accurate determination of the Curie point of nickel by.[26] They used a similar resistance technique to that described here but obtained superb results by paying considerable attention to sample preparation and temperature control. Around the transition temperature they took points at 20-mK intervals with temperature drift of $1\,\mathrm{Kh}^{-1}$. Zumsteg and Parks[26] argue that the Curie temperature is at the maximum of dR/dT. However, this is only true if the measurements of the (R, T) curve are taken at extremely small intervals of T. The mathematical method of calculating the gradient will ensure that dR/dT starts to decrease *before* the Curie point is reached as the temperature rises, since the five or seven points involved in gradient finding will 'bridge' the vital area. In other words the Curie point on the (R, T) curve is slightly higher in temperature than the maximum of the dR/dT curve. We feel that it is instructive for students to compare their own data with those obtained by Zumsteg and Parks with their more sophisticated equipment. However, we should emphasise that the intention of our experiment is not to produce extreme accuracy, but rather to devise an experiment that is instructive and interesting for undergraduate students.

The sudden change from ferromagnetism to paramagnetism in nickel is an example of a second-order phase transition. The study of phase transitions and critical phenomen has nowadays become very prominent (Stanley 1971, Hahne 1983). An important parameter in such studies is known as the reduced temperature ϵ which is given by

$$\epsilon = \frac{T - T_c}{T_c}$$

where T is the actual temperature and T_c the critical temperature, which in this case is the Curie temperature. All critical properties are proportional to the reduced temperature raised to some power. The problem of describing critical phenomena is to determine what that power is, i.e., to determine the values of the critical exponents. It can be shown[22,10] that the magnetisation $M(T)$ of a spin system is given by

$$M(T) \sim (-\epsilon)^{\beta},$$

where β is a critical exponent. From studies of the resistivity of ferromagnets, it can be shown[8] that just below the Curie temperature the magnetic resistivity R_{mag} varies as the square of the magnetisation and hence we can write

$$R_{\mathrm{mag}}(T) \sim (-\epsilon)^{2\beta}$$

Above the Curie temperature, our resistance data vary linearly with temperature. This is to be expected since the Debye temperature T_D for nickel is 154°C and hence we are measuring in a region where $T > T_D$. By extending to lower temperatures the straight-line portion of the (R, T) plot that is above the Curie temperature, it is possible to estimate the magnetic resistance by difference from the actual (R, T) line. A log-log plot then yields a value for β. The value obtained for β is obviously very sensitive to the exact value chosen for T_c. If we take $T_c = 358$°C, which is the generally accepted value,[16,24] we obtain $\beta = 0.47$. If, however, we take $T_c = 356$°C, which we believe is the best T_c from our (R, T) data, then we obtain $\beta = 0.40$. It would appear that the values obtained for β depend markedly on the measurement techniques as well as the state of the material. For nickel the literature quotes values ranging from 0.358[6] to 0.417.[19]

FIG.6 'Staring crowd' phenomenon - an analogy of a phase transition (for explanation see text)

5. Discussion

There are many examples of phase transitions occurring in nature. A well known one is the change from a gas to a liquid when the temperature is decreased below a certain value or alternatively the pressure is increased. A second example is the transition from paramagnetism to ferromagnetism as one moves below the Curie temperature. Others are found in binary liquid mixtures, binary alloys, antiferromagnetics, ferroelectrics, and crystals with structural phase transitions, liquid crystals, superconductors and superfluids.[22,10] The important feature of a phase transition is that the properties found in the normal phase are completely different from those observed in the condensed phase. If we consider the case of nickel, above the critical or Curie temperature T_c the system is in the normal paramagnetic phase. There is some short-range order that exists for nearby spins, causing them to tend to align, but no long-range order. However, below T_c, there is a long range order, which means that all the spins in the system spontaneously line up parallel to one another. This emergence of long-range order marks the transition to the ferromagnetic state. A humorous example of a phase transition which is similar to that just decribed is the 'staring crowd' phenomenon. This was first pointed out by Mattuck[18] and is illustrated in Fig. 6. In a crowd, people are looking in random directions and this is equivalent to the normal state. However, it sometimes happens that if one person starts staring at, for example, the tree shown in Fig. 6, a large number of other people in the crowd will also start staring there. This corresponds to the condensed phase.

The whole subject area of phase transitions and critical phenomena has become a major topic of condensed matter physics over the last twenty years. The historical background to this has been excellently summarised in the lucid article by Domb.[7] The major breakthrough in our understanding occurred in the early 1970s when K. G. Wilson proposed that the renormalisation-group technique could be relevant in studies of critical phenomena. This technique had been introduced in the 1950s in connection with the study of field theory. However, it has proved enormously fruitful for understanding a wide range of problems in condensed-matter physics. The renormalisation-group technique has been explained in simple terms by Wilson [25] where he discussed in some detail its application to the transition from paramagnetism to ferromagnetism when a ferromagnetic material is cooled to below its Curie temperature. Some idea of the importance of Wilson's work can be gauged by his award of the 1982 Nobel prize in physics.

Acknowledgements

We should like to thank Miss Susan Law for technical assistance and Miss Susan Fairhust for drawing Fig. 6.

References

1 N.W. Ashcroft and N.D. Mermin, *Solid State Physics*, Holt, Rinehart, and Winston (1976)
2 *Manual on the Use of Thermocouples in Temperature Measurements*, ASTM Spec. Tech. Publ. 470b, p.219 (1981)
3 G. Barnes, *The Physics Teacher* **24**, 86-87 (1986) and **24**, 204-210 (1986)
4 M. Blackman, *Contemp. Phys.* **24**, 319-31, (1983)
5 *British Standards BS4937 part 4*, British Standards Institution (1973)
6 J.D. Coken and T.R. Carver, *Phys. Rev. Lett.* **15**, 5330-5366 (1977)
7 C. Domb, *Contemp. Phys.* **26**, 49-72 (1985)
8 M.E. Fisher and J.S. Langer, *Phys. Rev. Lett.* **20**, 665-668 (1968)

9 J.N. Fox, N. Gaggini, and J.K. Eddy, *Am. J. Phys.* **54**, 723-26 (1986)
10 F.J.W. Hahne (ed.), *Critical Phenomena*, Lecture Notes in Physics 186, Springer (1983)
11 H.E. Hall, *Solid State Physics*, Wiley (1978)
12 J.C. Hopkins, to be published (1987)
13 C.M. Hurd, *Contemp. Phys.* **23**, 469-493 (1982)
14 R. Kamal, S. Sikri, and B.R. Sood, *Am. J. Phys.* **51**, 631-635 (1983)
15 M.P. Kawatra and J.I. Budnick, *Int. J. Magn.* **1**, 61-74 (1970)
16 G.W.C. Kaye and T.H. Laby, *Tables of Physical and Chemical Constants*, 15th ed., Longman, p.138
17 C. Kittel, *Introduction to Solid State Physics*, 6th ed., Wiley (1986)
18 R.D. Muttuck, *A Guide to Feynman Diagrams in the Many-Body Problem*, McGraw-Hill, p. 290 (1976)
19 J.L. Oddou, J. Berthier, and P. Penattot, *Phys. Lett.* **A45**, 445-446 (1973)
20 T.J. Quinn, *Temperatures*, Academic Press, p.402 (1983)
21 H.M. Rosenberg, *The Solid State*, 2nd ed., Oxford University Press (1982)
22 H. Stanley, *Introduction to Phase Transitions and Critical Phenomena*, (Oxford: Clarendon) 1971
23 M.B. Stearns, *Physics Today*, April 34-39 (1978)
24 R.C. Weast (ed), *Handbook of Chemistry and Physics*, 66th ed., Chemical Rubber Co., E 114 (1985)
25 K.G. Wilson *Sci. Am.* **241** (August) 140-157, (1979); *Rev. Mod. Phys*, **55**, 585-600 (1983)
26 F.C. Zumsteg and R.D. Parks, *Phys. Rev. Lett.* **24**, 520-4 (1970)

Resonance in coupled circuits

S.G. HAMMOND AND N.H. SAUNDERS
Department of Physics, University College of Swansea, Singleton Park, Swansea, SA2 8PP, UK

Introduction

The study of resonance in a simple series or parallel LCR circuit is included in most elementary courses on electric circuit theory. It is then often extended to the case of two such circuits, separately tuned to the same resonance frequency but coupled together in some way, commonly by the mutual inductance of the two coils.[1] Such a system is analytically straightforward and has interesting practical applications; the two identical resonance angular frequencies ω_0 of the non-interacting circuits are changed, one to a value higher than ω_0 and the other to a lower value, the change depending on the strength of coupling of the circuits now to be regarded as a single system. This well-known picture can be used as an electrical analog of energy level splitting in interacting atomic systems.[2] And the analogy is strengthened if it can be simply demonstrated that the coupling of additional LCR circuits leads to a corresponding increase in the number of resonances.

The experiment described is thus a qualitative demonstration of the resonance frequency spectrum for an electrical line comprising, in turn, one, two, three, ... series LCR circuits interacting via the mutual inductance of their coils. The apparatus used is to be found in most undergraduate laboratories and allows many obvious variations. In addition, it is interesting to consider the circuit behavior more quantitatively, using a variety of simplifying assumptions. This will be illustrated for the reasonably non-trivial case of four coils in line, the predicted resonance spectrum being compared with experimental data.

Apparatus and experimental details

Figure 1 shows a schematic diagram of the apparatus for the case of four coils. All of them were large flat coils about 30 cm in diameter, having about 400 turns which gave $L \sim 100\,\text{mH}$. They were individually tuned to about 5 kHz, using variable capacitance boxes. The audio frequency generator had a slow sweep facility usually set to cover the range $1 - 10\,\text{kHz}$ in $1 - 10\,\text{sec}$. To demonstrate the resonance frequency spectrum conveniently, a storage oscilloscope was connected as shown to sense the current in any one of the coils and was triggered from the audio signal generator. The mean distances between adjacent coils was approximately 5-10 cm, corresponding to reasonably tight coupling to reveal the otherwise degenerate frequency spectrum. Figure 2 shows a typical oscilloscope trace, where the negative portion has simply been suppressed by lowering the base line to the bottom of the screen.

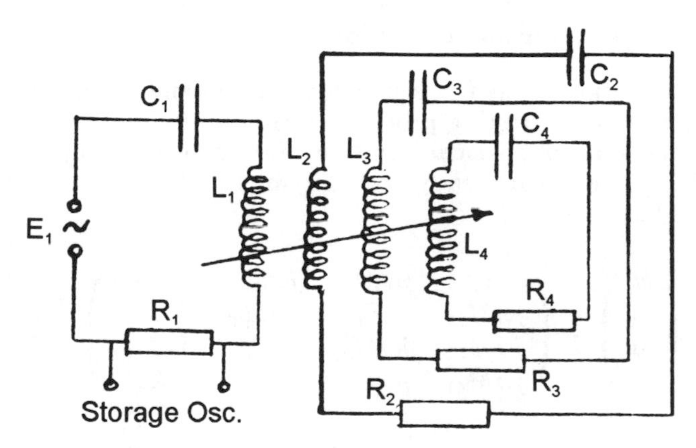

FIG.1 Circuit diagram for the case of four coils.

The changes in the pattern of resonances as first only one coil is used, and subsequently a second, third ... are added to the line, can easily be observed by incorporating switches in each of the LCR loops.

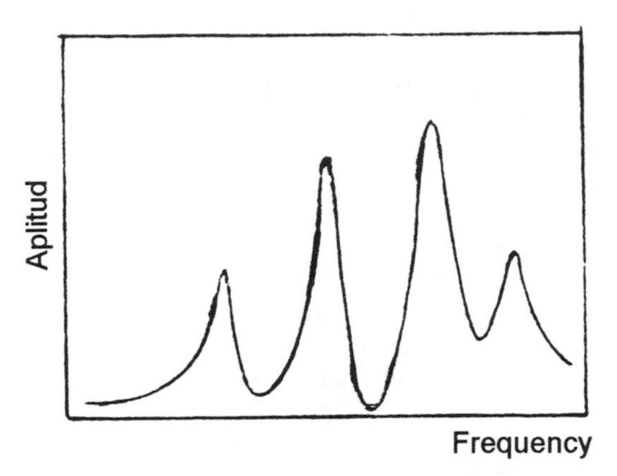

FIG.2 Typical resonance frequency spectrum for four strongly coupled coils.

Quantitative circuit behavior

The prediction of the frequency spectrum in any given coil configuration is an interesting problem in circuit analysis, which allows for some input of physical intuition. For a sinusoidal input voltage in Fig. 1 the linear steady state circuit equations can be written in the form

$$\begin{pmatrix} E_1 \\ 0 \\ 0 \\ 0 \end{pmatrix} = \begin{pmatrix} Z_1 & j\omega M_{12} & j\omega M_{13} & j\omega M_{14} \\ j\omega M_{21} & Z_2 & j\omega M_{23} & j\omega M_{24} \\ j\omega M_{31} & j\omega M_{32} & Z_3 & j\omega M_{34} \\ j\omega M_{41} & j\omega M_{42} & j\omega M_{43} & Z_4 \end{pmatrix} \begin{pmatrix} I_1 \\ I_2 \\ I_3 \\ I_4 \end{pmatrix}$$

where I_i are the loop currents and $Z_i = R_i + j\omega + 1/j\omega L_i C_i = R_i + j\omega L_i \left(1 - \omega_0^2/\omega^2\right) = R_i + j\omega L_i W$, ω being the angular frequency of the source and ω_0 the resonance angular frequency to which each of the loops is separately tuned. $M_{ij} = M_{ji}$ is the mutual inductance coefficient between coils i and j and is conventionally re-expressed as $k_{ij}\sqrt{L_i L_j}$, k_{ij} being the coupling coefficient ($0 \leq k_{ij} \leq 1$). In the case of tight coupling, it is often a reasonable approximation to take $R_i \ll \omega L_i W$, the resonant currents in this loss-free limit then tending to infinity. In terms of the matrix equation above, using Cramer's rule, current resonance will occur at those angular frequencies satisfying the following quartic in ω^2.

$$\begin{vmatrix} W & k_{12} & k_{13} & k_{14} \\ k_{21} & W & k_{23} & k_{24} \\ k_{31} & k_{32} & W & k_{34} \\ k_{41} & k_{42} & k_{43} & W \end{vmatrix} = 0$$

This equation was readily solved by computer successive interpolation (listing available on request) once values of the k_{ij} had been inserted. These values were calculated for the present case of co-axial flat coils from the expression $M_{ij} = k_{ij}.\sqrt{L_i L_j} = n_i n_j r N \, \mu H, n_i$ and n_j being the number of turns on the two coils concerned, r their common mean radius (in centimeters) and N a tabulated function depending on r and the distance apart of the coils.[3]

One of the coil configurations studied experimentally comprised coils 1, 3, and 4 in fixed positions and coil 2 moving over a range of about 14 cm between coils 1 and 3. The value of k_{34} (nearest neighbor interaction) was $\sim$ 0.45, k_{13} (next nearest neighbor) $\sim$ 0.12, and $k_{14} \sim$ 0.075. The value of k_{12} (or k_{23}) varied over the range 0.18-0.52. This configuration produced easily measured changes in the frequency spectrum, the actual resonance frequencies being read from a digital frequency meter across the manually adjusted audio oscillator.

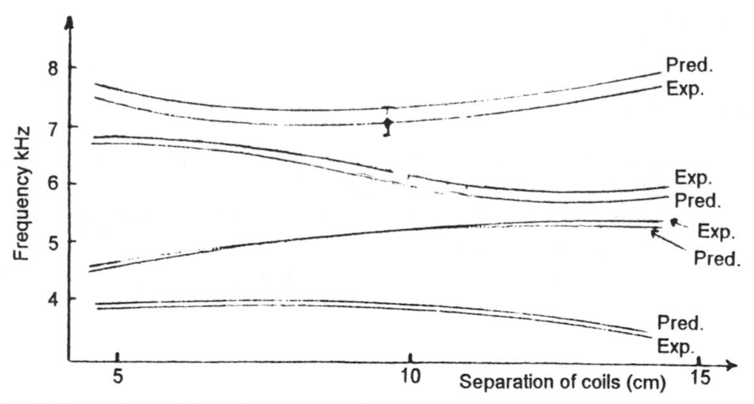

FIG.3 Experimental and predicted variation of the four resonance frequencies with distance between coils 1 and 2.

Figure 3 shows the four measured resonance frequencies as a function of separation between coils 1 and 2, together with the predicted variations for comparison. In most cases the agreement was reasonably good and within the estimated experimental error of ±2%. Similar conclusions were reached in the simpler cases of only two or three interacting coils. It is clearly possible to choose many different independent variables from the one illustrated here so the variety of possible experiments is extensive.

In addition it is intuitively attractive to consider what happens to the predicted frequency spectrum for a given coil configuration if all k_{ij} are set equal to zero except for the nearest neighbor coils. The effects of then adding-in next nearest neighbors, etc. can be easily studied, although this process is rather limited by the generally small number of coils involved. As a broad generalisation it was found that in all cases where well resolved resonance peaks could be observed, justifying the tight coupling approximation, agreement between experimental and theoretical frequency spectra became progressively worse as the supposed range of interaction decreased. This was particularly noticeable for the resonance of highest frequency, which was rather sensitive to the coupling assumptions made.

In using an experiment of this kind as an electrical analog to illustrate energy band formation in crystalline solids, it is interesting to extend it to consider parameters such as the overall bandwidth of the frequency spectrum and the significance of one or more of the LCR loops not being tuned to the same frequency as the rest of the loops in the line.

References

1 B.I. Bleaney and B. Bleaney, *Electricity and Magnetism*, Oxford university Press, chapter 9, (1957)

2 L. Brillouin. *Wave Propagation in Periodic Structures*, 2nd ed., Dover Publications (1953)

3 F.E. Terman, *Radio Engineers' Handbook*,p.72, McGraw Hill (1943)

Linear and nonlinear circuits

M. ELLIOTT AND D.G. SMITH
Department of Physics, University College of North Wales, Bangor, Gwynedd, LL57 2UW, UK

Abstract: Apparatus suitable for recording the responses of three simple circuits is described. The measured responses are compared with simple numerical solutions of the relevant circuit equations. The advantages of numerical methods, even when analytic solutions exist, are described.

Nonlinear circuits

The circuits shown in Figs. 1 and 2 have the silicon rectifier diode as their nonlinear element. The response of each circuit to the sudden application of a sinusoidal input voltage is recorded by means of a BBC-B microcomputer fitted with a low-cost digitial-storage-oscilloscope add-on unit. A signal generator provides a sinusoidal input voltage of a few volts amplitude at a frequency of about 10 Hz through a toggle switch. In practice, the oscilloscope sweep is initiated from the keyboard, and the switch is then moved from position (a) to position (b). This process is repeated a few times until the recorded trace shows that the input voltage was applied at the beginning of a signal cycle. Typical start-up waveforms are shown for each circuit.

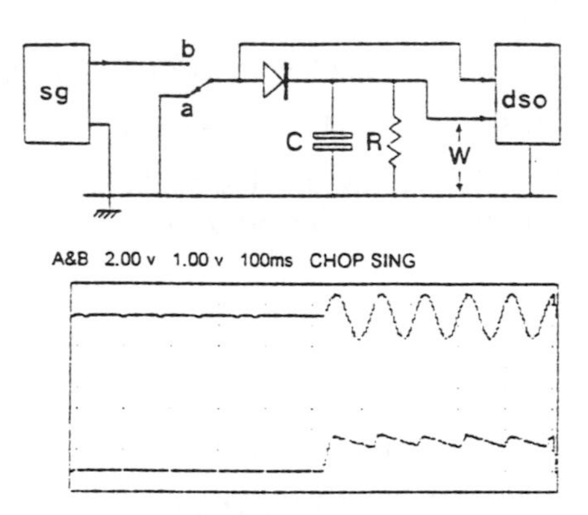

FIG.1 Half-wave rectifer circuit and measured start-up response

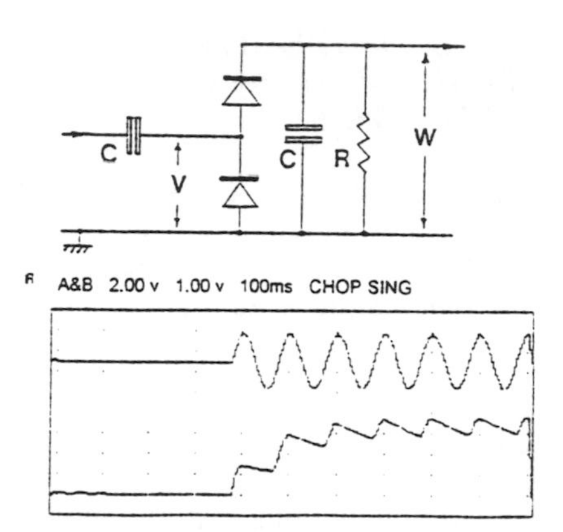

FIG.2 Voltage-doubler circuit and measured start-up response

The theoretical analysis of each circuit is straightforward once the $i-V$ characteristic of the silicon diode is known. The relationship $i = A\,[\exp(BV) - 1]$ describes this characteristic well. For the diodes we used, $A = 3 \times 10^{-10}$ A and $B = 24\,\mathrm{V}^{-1}$.

The circuit equation for the half-wave rectifier circuit is

$$dW/dt = A[\exp(B(U\sin(2\pi ft) - W)) - 1]/C - W/RC$$

The behavior of the voltage-doubler circuit is described by the two coupled equations

$$dV/dt = 2\pi fU\cos(2\pi ft) + A[\exp(-BV) - \exp(B(V - W))]/C$$
$$dW/dt = A[\exp(B(V - W)) - 1]/C - W/RC$$

These equations do not have analytic solutions in either case and must be solved numerically. This may be done very easily using a simple Euler solution obtained with a microcomputer. The Euler method is a first-order integration of the circuit equations. Taking the case of the voltage-doubler circuit, the solution value at time $t + \Delta t$ is obtained from the known value at time t in the following way

$$V + (dV/dt)\Delta t \to V$$

$$W + (dW/dt)\Delta t \to W$$

The solution obtained using this method is shown in Fig3. Some experimental data are marked on the graph and demonstrate the good quantiative agreement of the theoretical prediction.

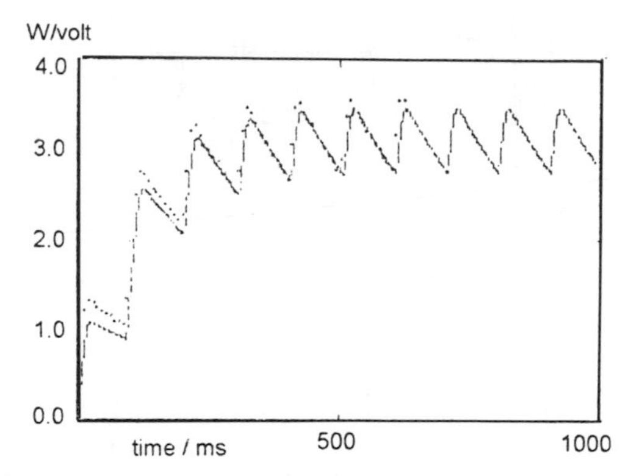

FIG.3 The voltage doubler: theoretical response (solid) and experimental data (circles)

Linear circuits

The circuit in Fig. 4 has a typical step-response as shown on the accompanying graph. Data may be recorded manually using a high-impedance voltmeter ($10^{10}\Omega$) and a stop-clock.

The circuit equations are

$$dV/dt = (U - V)R_1C - W/R_2C$$

$$dW/dt = (U - V)/R_1C - 2W/R_2C$$

These have analytic solutions of the form

$$W = U\left[1 + (2R_1/R_2)^2\right]^{-1/2} \exp(-\alpha t)[\exp(\beta t) - \exp(-\beta t)]$$

where

$$\alpha = [1 + 2R_1/R_2]\,/2R_1C$$

$$\beta = \left[1 + (2R_1/R_2)^2\right]^{1/2} /2R_1C$$

In practice, the derivation of these solutions and their quantitative display, even by a competent student, would be time-consuming (about a day's work probably), and it is doubtful if the necessary time would be set aside for this work in a course. However, the numerical solution of these equations using prepared software will allow the predictions of the circuit analysis to be displayed in a very short time (1 hour or less). Figure 5 shows the experimental data superimposed on three numerical solutions.

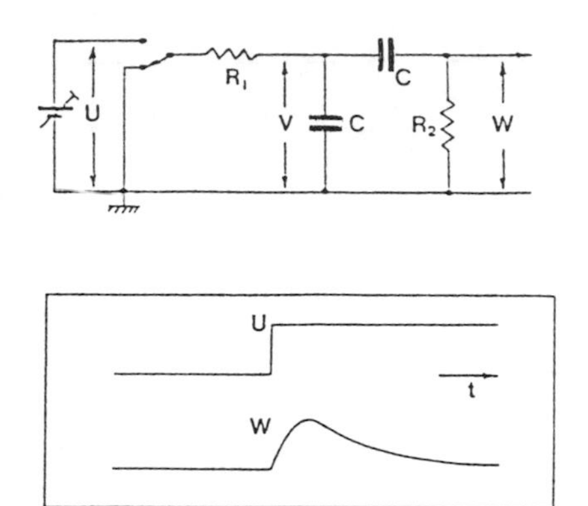

FIG.4 Simple linear circuit and typical step response

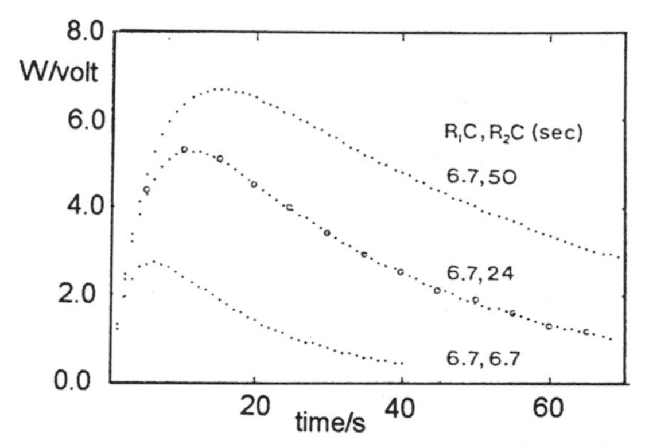

FIG.5 Simple linear circuit theoretical response (dotted) and experimental data (circles)

Summary

The theoretical analysis of simple circuits using unsophisticated numerical methods is powerful and time-efficient. In the case of nonlinear circuits numerical methods are usually essential. In the case of linear circuits, although numerical methods may not be essential, in practice they are quicker and easier to apply. This last fact means that the simple division of problems into those solved by calculus methods and those solved by numerical methods is not generally appropriate.

Fourier wave analysis using operational amplifiers

A.G. Bagnall
Harrow School, Harrow on the Hill, UK

The paper shows how three operational amplifiers can be used for simple analysis of a square and triangular wave signal. The experiment was designed as a part of the practical exam for the British Physics Olympiad. However, the idea, which is not claimed to be original, could be developed in several ways.

Three Phillip Harris 741 amplifiers are wired up as shown and this works as a sharp fixed frequency filter. The components can be chosen such that the circuit's natural period is around 500 Hz. It is quite well damped so it is not easy to see the oscillations. However, if connected to a suitable supply of sine, square, or triangle it will be seen to respond sharply.

To find the harmonics of the square wave it is only necessary to alter the signal generator such that the fundamental is now lower, e.g., f/3, so that the first harmonic now resonates the circuit. The relative amplitudes can be measured by the C.R.O., and are found to agree closely with theory. The accuracy does depend on the signal generator and how easily the signal generator can be set.

Copies of the question set are available.

Experiment II (90 minutes)

Apparatus

1 Three operational amplifiers (741)
2 One double beam C.R.O.
3 +15, 0, -15 V power supply
4 Two 1-nF capacitors
5 Resistors, $1\,\mathrm{k}\Omega$ and $10\,\mathrm{k}\omega$
6 Resistance box
7 Signal generator
8 A.C. voltmeter

Method

A. Wire up the amplifier as shown above (Fig. 1 and Fig. 2). The supply rails are $+15, 0, -15\,\mathrm{V}$.

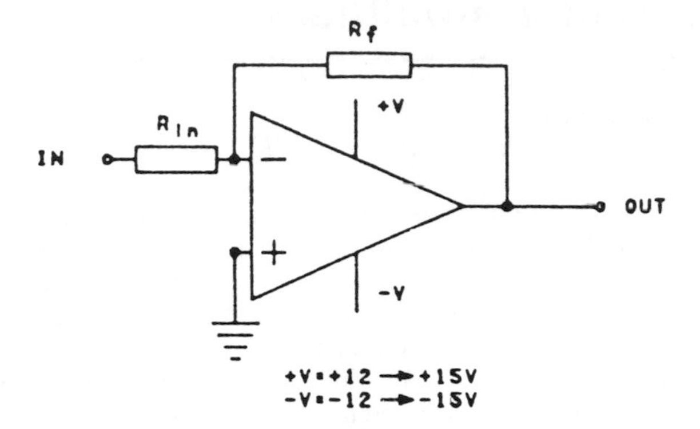

FIG.1

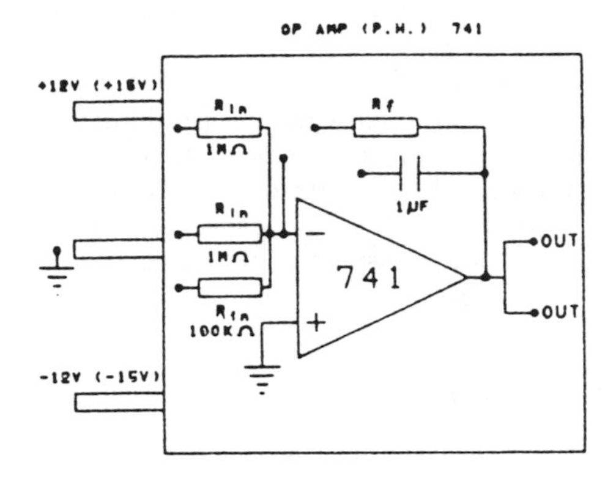

FIG.2

Theory shows that if the amplification factor A of the 741 is large and the 741 input resistance is larger, then

$$\frac{V_{\text{out}}}{V_{\text{in}}} = -\frac{R_f}{R_{\text{in}}}(\text{independent of } A)$$

Check this relationship for a fixed frequency and indicate the accuracy of your result.

Experimentally determine if the relationship is independent of frequency (a graph is not required). What is the purpose of the trim?

B. Replace the feedback resistor R_f by the $1\,\mu$F capacitor. If the 741 takes no current from the input then it follows that

$$V_{\text{out}} = -\frac{1}{CR}\int V_{\text{in}}\,dt$$

Justify this theoretically.

The circuit below (Fig. 3) should now behave as an integrator. You can check this experimentally, for a specific case, by connecting the input to +15 or −15 volts (into the 1MΩ input resistor).

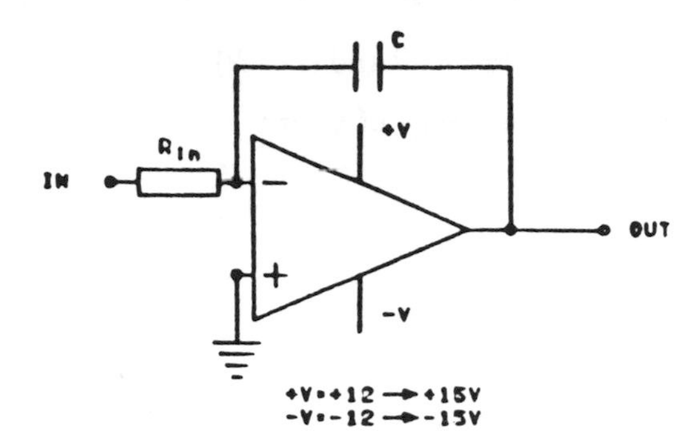

FIG.3

C. Show theoretically that three operational amplifiers can be used to solve the equation

$$\frac{d^2v}{dt^2} = -kv$$

Draw a circuit diagram and check that the circuit gives an oscillatory voltage v. [Hint: You will require a gain of ×10, and the trim may need adjusting.]

D. If you replace the 1 μF capacitors with 1 nF capacitors, the natural frequency of your circuit will be approximately 500 Hz. (You will have some difficulty in observing this because of the rapid damping of the circuit.) However, if a signal is fed into the circuit, it will resonate and its natural frequency may be observed. The signal is best fed from the signal generator using one of the other 1 MΩ resistors built into the operational amplifier box. You have now created a sharp filter whose frequency is approximately 500 Hz.

E. A square of triangular wave can be represented as a sum of a series of sine waves of different frequencies: e.g.,

$$v(t) = a_1 \sin(2\pi f_1 t) + a_2 \sin(2\pi f_2 t) + a_3 \sin(2\pi f_3 t) + \ldots..$$

Without altering your circuit, set as a fixed filter at approximately 500 Hz, find the following ratios for both triangular and square waves:

$$f_2/f_1 \text{ and } f_3/f_2$$

Without attempting to change the frequency of your filter, determine the ratios:

$$(a_2/a_1)^2 \text{ and } (a_3/a_2)^2$$

and comment on the accuracy of your results.

Characterization of pn junctions by microcomputer

S. Rakowski
Department of Physics, Staffordshire Polytechnic, Stoke-On-Trent, UK

Abstract: An undergraduate experiment is described that enables a student to investigate the forward and reverse characteristics of silicon and germanium pn junctions at various temperatures and to correlate the measured characteristics to those predicted theoretically by the Shockley equation. The experiment is controlled via a BBC microcomputer and is performed by the student in an interactive way.

Introduction

A standard, textbook analysis of the action of a semiconductor junction[1] yields the result that the current I flowing across a junction to which a voltage V is applied is given by

$$I = I_0 \left[\exp\left(\frac{eV}{kT}\right) - 1\right]$$

where I_0 is the reverse saturation current and T is the junction temperature. This is the original form of the Shockley equation and assumes that essentially all the current is derived from the diffusion of minority carriers through the junction. This assumption is reasonably valid for germanium and semiconductor materials with smaller energy gaps. In the case of wider energy gap materials such as silicon and gallium arsenide, however, another mechanism exists that significantly modifies the current. This derives from carriers that are generated and recombine in the depletion region.[2] When this effect is taken into account a modified version of the Shockley equation results $I = I_0 [\exp(eV/NkT) - 1]$ where N is an arbitrary constant whose value is primarily determined by the type of material used and other junction conditions.

The experiment

The practical measurement of the forward and reverse characteristics of a pn junction is not difficult. It can be achieved manually with a variable voltage power supply and accurate and sufficiently sensitive voltmeters and ammeters. In this experiment the measurements are performed automatically through a BBC microcomputer in a system shown in Fig. 1.

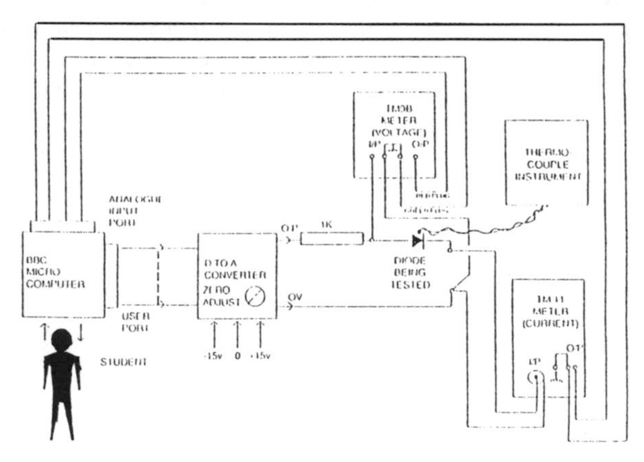

Fig.1 Interactive, computer-controled experiment investigating pn junction charateristics and the Shockley equation

The computer operates a digital-to-analogue converter via its user port. The output of the converter is applied to a buffer amplifier that produces the output voltage to be applied to the junction diode being tested. Software then generates a rising ramp of voltage at the output of the amplifier. The current flowing through the diode (I in the equation) and the voltage across it (V in the equation) are then measured by two standard laboratory multimeters that also produce output voltages proportional to the value being read on the meter. These voltages are read into the computer through two standard analogue input channels. In this way the computer measures a series of points of I against V for both the forward and reverse directions. The values are stored in arrays for subsequent analysis. The student can observe the readings being taken by the computer by observing the two multimeters. A thermocouple attached to the diode being tested is used to measure the junction temperature and the student must record this value for later, keyboard input to the computer.

The measured and stored values are then output in a tabular

form, on a printer for the two materials tested (i.e., silicon and germanium). The reverse characteristic of each material is then plotted from these results and values of I_0 are estimated from the plots for the two materials at the measured temperature. Typical plots obtained are shown in Fig. 2.

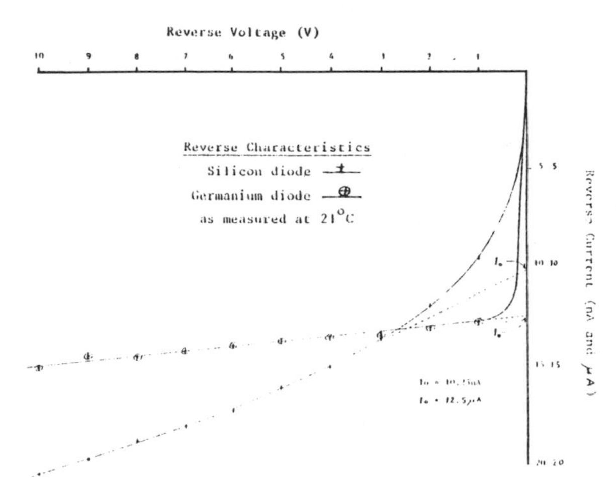

FIG.2 Reverse characteristics

The estimated values of I_0 are then input to the computer by the student and the computer generates a prediction of the forward characteristics from the original form of the Shockley equation (i.e., for $N = 1$). Next the student uses the computer to carry out a series of iterations to perform a least squares fit of the measured characteristic to the Shockley equation prediction in order to obtain an optimum value of N giving the best fit. The modified Shockley equation is then used by the computer to generate a new prediction of the forward characteristics, using these optimal values of N. Finally, comparative plots of the characteristics can be made from these results for the two materials. These are shown in Figs. 3 and 4.

The complete experiment can be easily repeated at a higher junction temperature in order to demonstrate the effect of temperature on the junction characteristics. This is simply achieved by blowing hot air over the diodes as they are tested. The equivalent results obtained at a higher temperature are shown in Figs. 5, 6, and 7. An error analysis between measured values and optimal Shockley predictions is also obtained from the computer.

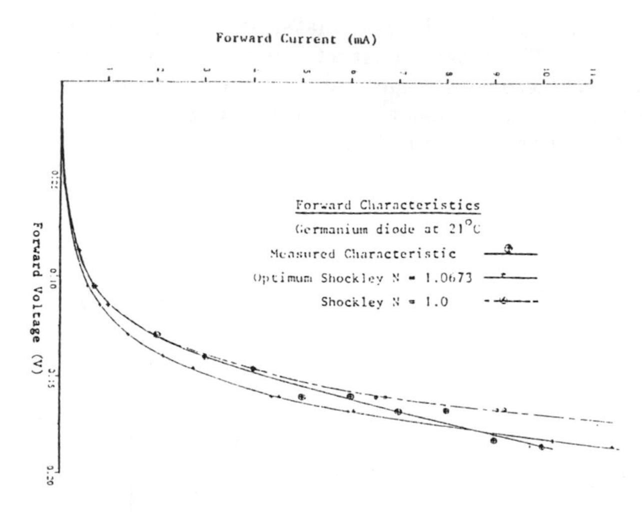

FIG.3 Forward characteristics

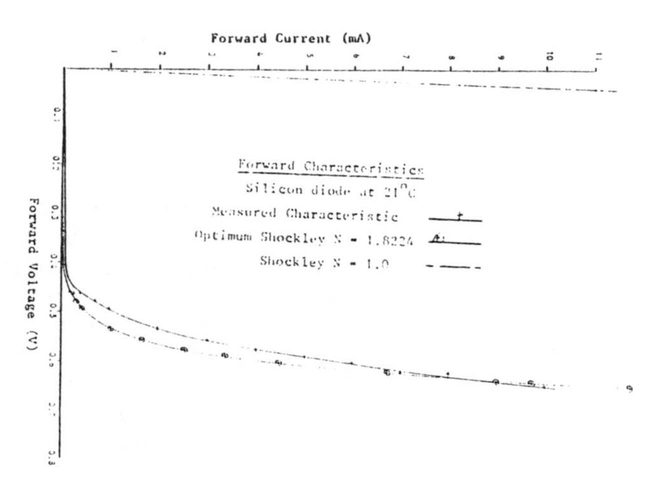

FIG.4 Forward characteristics

Conclusions

The inclusion of a microcomputer, in what is a simple experiment to perform manually, has many advantages. It enables many more readings to be taken more accurately and in a more repeatable fashion. This in turn enables the student to compare the characteristics of different materials, at different temperatures, in a rel-

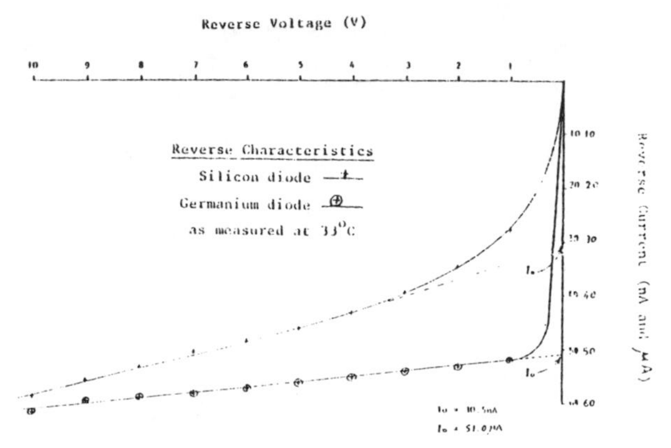

FIG.5 Reverse characteristics

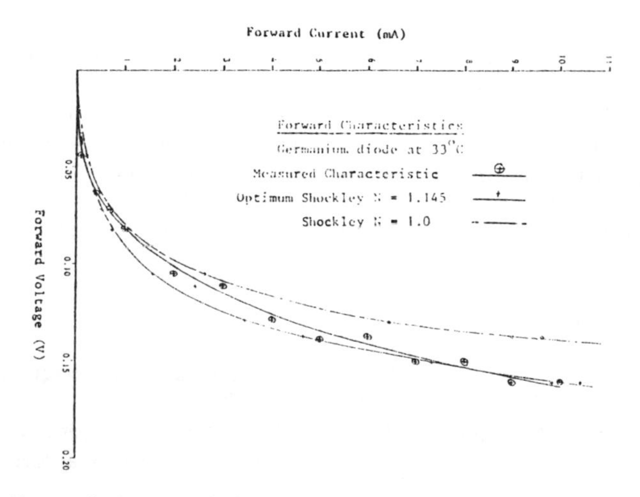

FIG.6 Forward characteristics

atively short time (the experiment can be comfortably completed within a three-hour laboratory session). Furthermore, the powerful, automatic computational facilities of the computer allow the student to easily calculate an optimum value for N in the Shockley equation for the two materials tested under the different operating conditions. This is difficult to achieve manually. During the course of this computation, which is an interactive procedure between student and computer, the student is also made aware of the limits of accuracy of original data by the attainment of a minimum

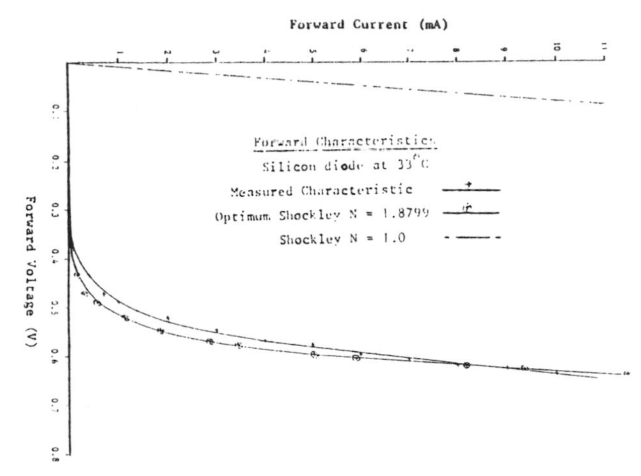

FIG.7 Forward characteristics

sum of the squares of the errors (ΣE_r^2) that is achieved after about two decimal places have been generated in the optimal value of N.

In conclusion, the claim can be made that the automation of this experiment allows the student to concentrate on the all-important physical principles involved in the phenomena under investigation, but it also provides a clear demonstration of the powerful data-handling facilities provided by the inclusion of a microcomputer.

References

1 D.H. Navon, *Semiconductor Micro-devices and Materials*, Holt, Rinehart, and Winston, pp.120-125, 1986

2 C.T. Sah, R.N. Noyce, and W. Shockley, 'Carrier generation and recombination in pn junctions and pn junction characteristics', Proc. I.T.E. **45**, 1228 (1957)

Physics simulations for the Advanced Level Course

A.G. BAGNALL
Harrow School, Harrow on the Hill, UK

These demonstrations are designed to support Advanced Level Physics Courses, and in particular the Nuffield course. First year undergraduates studying physics and related subjects may find these programs useful. The topics chosen are those that the author feels are difficult to demonstrate practically or to illustrate conventially.

MENU 1

Diffraction experiments

The idea here is to show how diffraction patterns are built up from individual photons to give a picture that is similar to that which can be seen with a red laser or white light. The target is rotated through 90° for visibility, but students do not seem disturbed by this rotation. If left on for too long, nearly all points on the target will be hit and since each pixel is either on or off the correct graduation of intensity will be lost. The removal of pixels at a steady rate would be possible, but this would slow the program down.

F1 One slit. Red photons are shot at the target. Pressing ENTER changes the color to blue and then to green. Lastly white is represented by blue, green, and red photons. In practice the photons are associated with a continuous change of color, energy, and wavelength.

F2 Two slits. This is similar to 'oneslit' except that the user is given a choice of slit separation and slit size. The program initially runs slowly to show the optical paths. Then pressing RETURN speeds up the process.

F3 One hole. This shows the pattern formed by a small hole.

F4 Two slits (white). Similar but fires red, blue and green photons.

Rutherford scattering

F5. The red marker simulates the behavior of an alpha particle (He^{++}) when it meets a gold nucleus.

It is assumed that (i) the gold nucleus is very massive, (ii) classical Physics holds and the alpha particle speeds are not relativistic, and (iii) the firings are at random, but no account is taken of target areas, so that the chances of a head-on collision are high whereas in reality such collisions are very rare.

The numerical integration is

$$S' = s + vDt$$

$$v' = v + aDt$$

where a is given by F/m and F obeys the inverse square law

$$F = qq'/4\pi r^2$$

It may be of interest to note that 5 000 of these calculations are performed per plotted point to give accuracy. This part of the program was written directly in 8086 code, using integer arithmetic.

Gravitation

F6. This uses similar mathematical calculations to the Rutherford scattering. But even though a large number of calculations are performed errors creep in, especially where the acceleration is high.

The speed of the satellite is asked for in km/s (the escape velocity from the Earth is just over 11 km/s).

The starting position is also asked for in km from the center of the Earth. It is interesting to note that to achieve a circular orbit values must be chosen with care.

The equations are

$$GMn/r^2$$

i.e.,

$$v = k/r$$

F7/8. A simple one-dimensional wave equation is solved numerically. The total energy is chosen such that the first solution is obtained at an energy of 20.4×10^{-19} J.

First a table is displayed showing the potential energy, the kinetic energy, the amplitude A of the wave function, and the square of the amplitude I. The relative probability of finding an electron at a point is proportional to I and this is used for the final display of the density distribution.

The total energy in this case is chosen to be near the second solution. On pressing ENTER a new value is tried that is nearer: the A and I curves are not asymptotic, but as ENTER is pressed it can be seen that successive solutions are nearer because the A and I curves approach the r axis.

The 'correct' solution is passed and then displayed again. Finally, a density distribution is given as before.

MENU 2

Experiments on quanta

These are designed to illustrate the distribution of energy quanta in an Einstein solid. The random quanta are assumed to be of equal energy and able to move freely throughout the solid.

The green squares represent atoms (oscillators), the red bars quanta. An atom is chosen at random, a quantum removed; then another atom is chosen at random, and the quantum transferred to it.

For large numbers it can be shown that $NI/(NI+1) = 1+N/n$ where N is the total number of atoms and n is the number of quanta. The numbers shown here are not large, but they illustrate the argument.

F1 Cold. In the cold program the movement of hot to cold can be seen since the right side of the frame of atoms has initially no quanta.

5 000 moves are made, but for speed the histogram is not displayed after 300 moves until the end of 5 000 moves.

F3 Hot. In the hot program $N = n = 400$ and in the cold $N = 400, n = 200$.

The new quanta versions are in line with the revised Nuffield course requirements and are on function keys as follows:

F2 New hot
F5 New cold
F7 New warm

Charged particles in a magnetic field

Two programs are given.

F4. In this case, the viewing angle is asked for, the angle that the particles initial direction makes with the field, the value of B, and the initial speed.

F6. Negative and positive charges in a magnetic field. B and V are asked for. Suitable values are 8×10^{-2} T and 4×10^{7} m/s. The relevant equation is $Bqv = mv^2/r$.

Different views of the helix are given. (The body has a velocity component parallel to the magnetic field).

Moving clock

F8. This is Einstein's 'moving light clock'

The green frame represents a cylinder with perfect mirrors at each end. Light bounces back and forth between the two. Einstein assumed that the speed of light (*in vacuo*) was independent of the movement of the source or observer. (Light from the Andromeda galaxy moving relative to us at some 600 km/s moves at the same speed in the laboratry). If the green frame moves past an observer, he or she sees the light go further (the yellow line) than the observer who moves with the frame (the green line).

Speeds are the same – distances are different, i.e., the times must be different $t' = t/[1 - (v/c)^2]^{1/2}$

MENU 3

Decay

This is a simple simulation of radioactive decay. The 'green' atoms decay into the 'red' atoms which, in turn, decay into the stable 'yellow' atoms. Graphs are plotted showing the number of atoms left at any time. Three programs are available.

F1. Equal half-lives

F2. Red to yellow half the half-life of green to red.

F3. Green to red half the half-life of red to yellow.

F4. Input half-lives.

Molecular dynamics simulation

C. Isenberg , R.F. Fowler, and M.L. Williams
Department of Physics, University of Kent at Canterbury, Canterbury, Kent, UK

Introduction

This computer 'experiment' was set for the 1986 International Physics Olympiad held at Harrow School. RML Nimbus microcomputers had been programed to solve the Newtonian equations of motion for interacting particles in a two-dimensional box. Information concerning the system, at discrete time steps, was available to the students at the depression of a key. They were not required to write or understand the program loaded into the Nimbus. The program was used to test the students' understanding of the laws of mechanics, statistical mechanical concepts, and thermodynamics. Details of the program and the numerical techniques can be referred to the papers and books given in the references. Copies of the program, on disc, can be obtained from the authors.

Apparatus

1. RML Nimbus Microcomputer
2. Ten sheets of graph paper

Information

The microcomputer has been programed to solve the Newtonian equations of motion for a two dimensional system of 25 interacting particles in the x, y plane. It is able to generate the positions and velocities of all particles at discrete, equally spaced, time intervals. By depressing appropriate keys, which will be described, access to dynamical information about the system can be obtained.

The system of particles is confined to a box that is initially, at time $t = 0$, arranged in a two dimensional square lattice array. A picture of the system is displayed on the screen together with the numerical data requested. All particles are identical; the colours are used to enable the particles to be distinguished. As the system evolves in time the positions and velocities of the particles will change. If a particle is seen to leave the box the program automatically generates a new particle that enters the box at the opposite face with the same velocity; thus conserving the number of particles in the box.

Any two particles, i and j, separated by a distance r_{ij}, interact with a potential

$$U_{ij} = 4\epsilon \left[\left(\frac{\sigma}{r_{ij}} \right)^{12} - \left(\frac{\sigma}{r_{ij}} \right)^{6} \right]$$

σ is a constant length characteristic of the potential and ϵ is a constant with the dimensions of energy.

It is convenient to use reduced dimensionless units throughout the computation. The reduced units given below are used throughout the calculations, where m is the mass of each particle, V is a velocity, t a time, E an energy, and T is a temperature.

NAME OF UNIT	SYMBOL	SYMBOLIC FORM
Reduced Distance	r^*	$r^* = r/\sigma$
Reduced Velocity	V^*	$V^* = V\,[m/(48\epsilon)]^{+1/2}$
Reduced Time	t^*	$t^* = t\,[m\sigma^2/(48\epsilon)]^{-1/2}$
Reduced Energy	E^*	$E^* = E/\epsilon$
Reduced Mass	m^*	$m^* = 48$
Reduced U_{ij}	U^*_{ij}	$U^*_{ij} = U_{ij}/\epsilon$
		$= e\left[(r^*_{ij})^{-12} - (r^*_{ij})^{-6}\right]$
Reduced Temperature	T^*	$T^* = kT/\epsilon$

The numerical data displayed on the screen must be multiplied by the appropriate unit, or power of these units, to give the correct, dimensional, form.

Instructions

The computer program allows one to access three distinct sets of numerical information and display them on the screen. Access is controlled by the grey function keys on the lefthand side of the keyboard labelled F1, F2, F3, F4, and F10. The keys should be just pressed and released: do not *hold down* a key or press it repeatedly. The program may take up to 1 s to respond.

First information set: Problems 1–5

Depressing F1 allows the screen to display the three quantities $\langle VX, n\rangle$, $\langle VY, n\rangle$ and $\langle U\rangle$ defined by

$$\langle VX, n\rangle = \frac{1}{25}\sum_{i=1}^{25}(V_{ix}^*)^n$$

$$\langle VY, n\rangle = \frac{1}{25}\sum_{i=1}^{25}(V_{iy}^*)^n$$

and

$$\langle U\rangle = \frac{1}{25}\sum_{i=1}^{25}\sum_{j=1}^{25}U_{ij}^*$$

where V_{ix}^* is the dimensionless x-component of velocity for the ith particle, V_{iy}^* is the dimensionless y-component of velocity for the ith particle, and n is an integer with $n \geq 1$ (note that the summation over U_{ij}^* excludes the cases in which $i = j$).

After depressing F1 it is necessary to input the integer n $(n \geq 1)$ by depressing one of the white keys, in the top row of the keyboard labelled with an integer, before the information appears on the screen.

The information is displayed in dimensionless time intervals Δt^* at dimensionless times

$$S\Delta t^* \qquad (S = 0, 1, 2, \ldots)$$

Δt^* is set by the computer program to the value $\Delta t^* = 0.100000$. The value of S is displayed at the bottom righthand side of the screen. Initially it has the value $S = 0$. The word ‘waiting’ on the screen indicates that the calculation has halted and information concerning the value of S is displayed.

Depressing the long bar, the 'space bar', at the bottom of the keyboard will allow the calculation of the evolution of the system to proceed in time steps Δt^*. The current value of S is always displayed on the screen. Whilst the calculation is proceeding the word "running" is displayed on the screen.

Depressing F1 again will stop the calculation at the time integer, indicated by S on the screen, and display the current values of

$$\langle VX, n\rangle$$

$$\langle VY, n\rangle$$

and

$$\langle U\rangle$$

after depressing the integer n. The evolution of the system continues on pressing the long bar.

The system can, if required, be reset to its initial state at $S = 0$ by pressing F10 twice.

Second information set: Problem 6

Depressing F2 *initiates* the computer program for the compilation of the histogram in Problem 6. This program generates a histogram table of the number, ΔN, of particle velocity components as a function of dimensionless velocity. The dimensionless velocity components, V_x^* and V_y^*, are referred to collectively by V_c^*. The dimensionless velocity range is divided into equal intervals $\Delta V_c^* = 0.05$. The centres of the dimensionless velocity "bins" have magnitudes

$$V_c^* = B\Delta V_c^* \qquad (B = 0, \pm 1, \pm 2, \ldots)$$

When the long bar on the keyboard is pressed the 2×25 dimensionless velocity components are calculated at the current time step, and the program adds one, for each velocity component, into the appropriate velocity 'bin'. This process is continued, for each time step, until F3 is depressed. Once F3 is depressed, the accumulated, histogram table is displayed. The accumulation of counts can be continued by pressing the long bar. (Alternatively, should

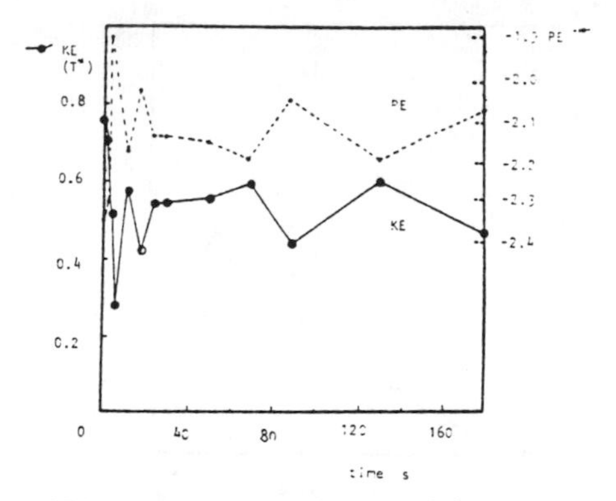

FIG.1 Variation of kinetic energy and potential energy

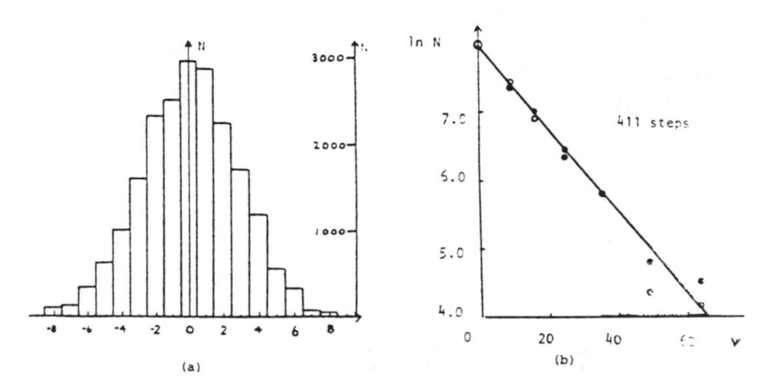

FIG.2 Maxwell – Boltzman distribution

one wish to return to the initial situation, in which the program is *initiated* with zero in all 'bins', one can press F2.)

The accumulation of histogram data should continue for about 200 time steps after initiation.

In the thermodynamic equilibrium the histogram can be approximated by the relation

$$\Delta N = A\exp\left[-24\left(V_c^*\right)^2/\alpha\right],$$

where α is a constant, associated with the temperature of the system, and A depends on the total number of accumulated velocity components.

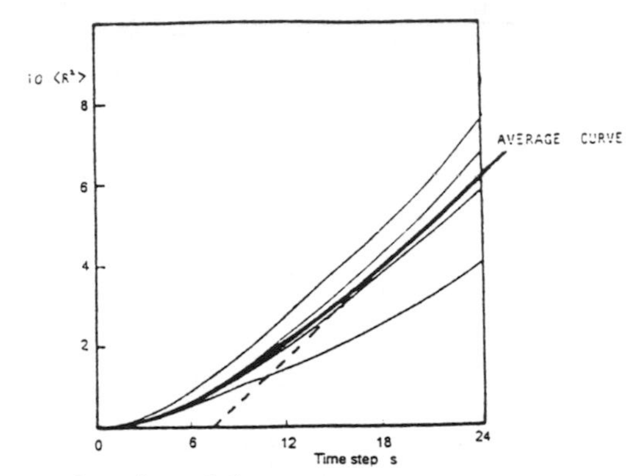

FIG.3 $\langle R^2 \rangle$ as a function of time

Third information set: Problem 7

Depressing F4 followed by the long bar at any time during the evolution of the system will initiate the program for Problem 7. The program will take some 30 seconds, in real time, before displaying a table containing the two quantities

$$\langle RX, 2\rangle = \frac{1}{25}\sum_{i=1}^{25}[x_i^*(S) - x_i^*(SR)]^2$$

and

$$\langle RY, 2\rangle = \frac{1}{25}\sum_{i=1}^{25}[y_i^*(S) - y_i^*(SR)]^2$$

where x_i^* and y_i^* are the dimensionless position components for the i'th particle. S is the integer time unit and SR is the fixed initial integer time at which the program is initiated by depressing F4. It is convenient to introduce integer

$$SZ = S - SR.$$

The program displays a table of $\langle RX, 2\rangle$ and $\langle RY, 2\rangle$ for

$$SZ = 0, 2, 4 \ldots 24$$

Prior to the display appearing on the screen a notice 'Running' will appear on the screen indicating that a computation is proceeding.

Depressing F4 followed by the long bar again will initiate a new table with SR advanced by 24 units.

Computation problems

1. Verify that the total momentum of the system is conserved for the times given by

$$S = 0, 40, 80, 120, 160$$

State the accuracy of the computer calculation.

2. Plot the variation in kinetic energy of the system with time using the time sequence.

$$S = 0, 2, 4, 6, 12, 18, 24, 30, 50, 70, 90, 130, 180$$

3. Plot the variation in potential energy of the system with time using the time sequence in 2.

4. Obtain the total energy of the system at the times indicated in 2. Does the system conserve energy? State the accuracy of the total energy calculation.

5. The system is initially, at $S = 0$, *not* in thermodynamic equilibrium. After a period of time the system reaches thermodynamic equilibrium in which the total dimensionless kinetic energy fluctuates about a mean value of T^*. Determine this value of T^* and indicate the time, SD, after which the system is in thermodynamic equilibrium.

6. Using the accumulated velocity data, during thermodynamic equilibrium, draw up a histogram giving the number ΔN of velocity components against reduced velocity component, using the constant velocity component interval $\Delta V_c^* = 0.05$, specified in the table available from the SECOND INFORMATION SET. Data accumulated from approximately 200 time steps should be used and the starting time integer $S = SDD$ indicated.

Verify that ΔN satisfies the relation

$$\Delta N = A\exp\left[-24\left(V_c^*\right)^2/\alpha\right]$$

where α and A are constants. Determine the value of α.

7. For the system of particles in thermodynamic equilibrium evaluate the average value of $R^2, < R^2 >$, where R is the straight line distance between the position of a particle at a fixed initial time number SR and time number S. The time number difference $SZ = (S - SR)$ takes the values

$$SZ = 0, 2, 4, ...24$$

Plot $\langle R^2 \rangle$ against SZ for any appropriate value of SR. Calculate the gradient of the function in the linear region and specify the time number range for which this gradient is valid.

In order to improve the accuracy of the plot repeat the previous calculations for three, additional, different values of SR and determine $\langle R^2 \rangle$ for the four sets of results together with the 'linear' gradient and time number range.

Deduce, with appropriate reasoning, the thermodynamic equilibrium state of the system, either solid or liquid.

References

1 A. Rahman, *Phys. Rev.* **A136**, 405 (1964)
2 L. Verlet *Phys. Rev.* **A159**, 98 (1967)
3 D. Nicholson and N.G. Parsonage, *Computer Simulation and Sattistical Mechanics of Adsorption*, Academic Press, New York, 1982
4 J.P. Hanson, and I.R. MacDonald, *The Theory of Simple Liquids*, Academic Press, 1976

Computer-aided draughting and analysis of electronic circuits

N.B. CRYER
Royal Holloway and Bedford New College, Egham, Surrey, UK

Abstract: **This report outlines the advantages of using computer simulations in electonrics as a partial replacement for laboratory work. It explains the use of one such package and describes how it functions. The report concludes with plans for further developments.**

The advantage of simulation in electronics

Electronics is a very practical subject. It is normally taught with the expectation that students should be able to design and construct working circuits. It is natural, therefore, to want to include a large amount of practical work in the teaching. It is, however, very time consuming to run practical classes.

At Royal Holloway and Bedford New College we have tried using computer simulations as a replacement for part of the electronics practical work. Whereas circuit building and complete analysis in a laboratory would take at least three hours for the most able students and more like a full day for the less dextrous, all students using our computer simulation can enter a circuit and see a complete analysis in less than half an hour.

This report starts by outlining the advantages of using simulations. It explains how the user draws a circuit and describes the nature of the analysis and the form of the presentation. It then describes how the simulation functions. The report concludes with plans for further developments of the simulation.

There are a number of important advantages associated with the use of computer simulations in the context of electronics:

(i) The handling of circuits via the conventional circuit diagrams drawn on the screen conforms more naturally with ways of thinking about circuits than the mess of wires that are normally associated with any practical exercise. This means that the simulation should be a better tool for learning about and understanding circuit properties.

(ii) Ease of editing component values provides a facility whereby the students may readily experiment and thereby learn in a more natural way. Such experimentation in the lab by trying different component values is rarely followed through by seeing the full significance on the complete analysis of the circuit.

(iii) There is no longer the need to provide the same degree of close supervision as is associated with laboratory work.

(iv) The computers can be kept outside the laboratory thereby providing wider access over an extended range of hours.

Using the simulation

To draw a circuit the user moves a cursor on the screen to where a component is required. He or she then presses a key such as *R* for resistor or *C* for capacitor to select the component to be drawn there. The image then appears on the screen. The user is then asked to select a value for the component. A circuit of almost any complexity may built up limited only by the resolution of the screen image and a maximum of 40 nodes (junction points) in the circuit. A circuit diagram may be saved, recalled and edited. A hard copy of the circuit may be obtained at any time on a standard dot matrix printer attached to the computer. An example of a typical circuit drawn by a student is shown in Fig. 1.

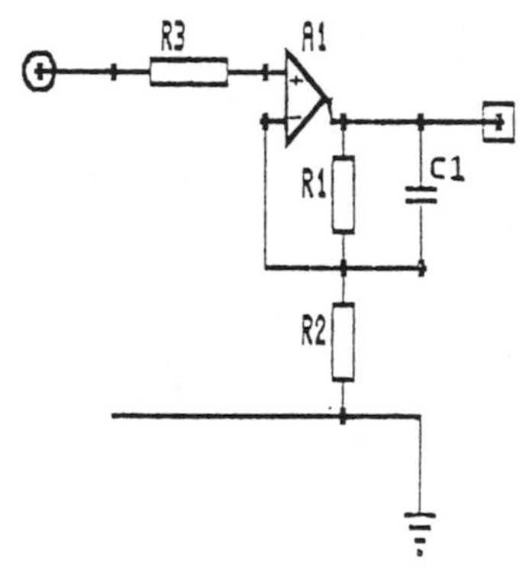

FIG.1 Operational amplifier circuit

A completed circuit may be analysed for gain, phase relation and impedance. Figure 2 shows part of an analysis for the circuit shown in Fig. 1. The range of frequencies over which the analysis takes place is 1Hz to 100 kHz. This range is an arbitrary choice and is simple to change. The wide range of frequency and gain can only be accommodated on a graph by using log scales. Because of the limited resolution and range of colors available on the IBM screen modes available to us, it is impossible to get as much detail on the screen as we would like when presenting the analysis. One limitation has been that we could not work out a way of including a label for the units along the frequency axis, although this has been added by hand for Fig. 2.

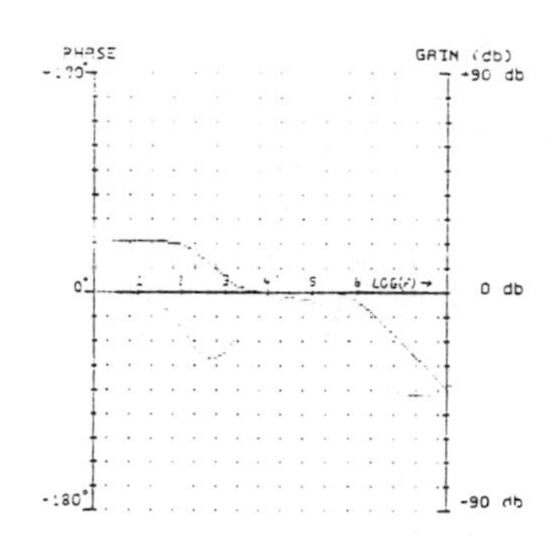

FIG.2 Part of the analysis of an operational amplifier circuit.

The simulation has been written in Turbo Pascal. This particular version of the language suffers from severe memory limitations, which have been a problem for us. We have found it necessary to write the program in two modules to overcome the language deficiencies. The user is, however, unaware of the modularisation of the program. The simulation should run on most IBM PC look alikes that use an EGA, CGA, or Hercules card, and also on the RML Nimbus Computer. The latter computer is used in a large number of London schools, one of which has done field trials for us.

The calculations involved in the analysis of the circuit require that a circuit containing n nodes be represented by an $n \times n$ array holding the admittance between every node and every other node. In programming terms this admittance matrix is simple to fill. Each time a component is added to the circuit diagram the program numbers the nodes joined by the component. The admittance caused by the component will be known as soon as its value has been specified by the user. As the program will also know at this stage the nodes between which the component has been drawn it can fill in this part of the admittance matrix. Once the student has completed the circuit drawing the complete admittance matrix representing the circuit will also be available.

The analysis requires the mathematical reduction of this $n \times n$ matrix to a 2×2 representing just the input and output nodes. Standard expressions then give the gain from input to output, the phase relation between the input and the output as well as the impedance as measured at the input and output.

Plans for the future

At present although the simulation can calculate the steady dc voltages around a circuit it cannot yet display these. Such a display would allow the bias conditions for any circuit to be examined, which is particularly important for transistor circuits. This is planned in the next development.

Another intention is to add the facility for the user to apply a waveform at the input of the circuit and to examine it at any point in the circuit using a simulation of an oscilloscope display synchronised to the input waveform. This will show dc saturation effects as well as any distortion due to the frequency and phase response. The product will provide more complete simulation of the true laboratory experience.

There are no plans to drop electronics practicals completely at Royal Holloway and Bedford New College as it is felt important for students to handle devices, wire them into circuits, and connect test equipment. This then has to be set to the correct ranges and readings interpreted in the light of these settings – all factors not covered in the simulation.

Computer-aided studies of nonlinear and chaotic systems

J. SWAGE*, G.J. KEELER, B.W. JAMES,
A.D. BOARDMAN, AND G.S. COOPER**
University of Salford, Salford, M5 4WT, UK

Abstract: The idea that given nearly 'exact' initial conditions governing the behavior of a particular physical system we should be able to predict its evolution indefinitely is false. In certain situations, if the starting parameters differ even only infinitesimally, the ensuing behavior of each system can produce totally dissimilar outcomes. The systems are then said to behave in a chaotic manner. However, the use of computer simulations of such systems allow the initial conditions to be set *exactly*, and so the 'chaotic' behavior is reproducible and deterministic. In this way sense can be made of the random evolution.

Chaotic phenomena

The Software Development Unit in the Department of Pure and Applied Physics at the University of Salford has been set up on the basis of a grant awarded by the Computers in Teaching Initiative (CTI). It has developed high quality computersimulations for teaching undergraduate students physics and computational physics. The software is written in strict, standard FORTRAN77 and uses the device independent Graphical Kernel System (GKS) for the graphical input/output. Some of the programs have been collected together to form a software package on nonlinear and chaotic systems.[1]

Many nonlinear systems, such as nonlinear resonant electrical circuits, nonlinear optical systems and the driven, damped simple and spherical pendula display very interesting properties for certain ranges of parameter values. Such effects include spontaneous oscillations that undergo period doubling leading ultimately, after intervals of intermittent behavior, to chaotic motion. See reference 2 for background reading to the subject.

* NNC Ltd., Knutsford

** Information Technology Institute, University of Salford

Mathematical model

The universal nature of such effects can be brought out by the use of computer experiments, and an understanding of the mechanisms involved can be gained by the study of a simple but very effective numerical model. A simple mathematical system that exhibits many of these effects in dramatic detail is the one-dimensional logistic map. This is an iterative process that takes the form of the mapping

$$x_1 = f(x_0)$$

$$\cdots\cdot$$

$$x_i = f(x_{i-1})$$

In this procedure the iteration number i takes the place of time and x_0 is the seed. The nth element in the sequence is

$$x_n = f(f(..f(f(x_0))..))$$

$$= f^n(x_0)$$

where n is the total number of applications of f [$f^n(x)$ is not the nth power of $f(x)$; it is the nth iterate of f].

Programs 1 and 2 in the software package[1] study the system when the function $f(x)$ is defined to be

$$f(x) = \alpha x(1 - x)$$

For certain values of α, the iterative procedure settles down to a self-repeating point, called a (stable) fixed point. There also exists an unstable fixed point (at $x = 0$) that repels the iterates no matter how close x_0 is to it, unless x_0 is exactly at the top of the hill it will always roll down to the stable equilibrium position at the bottom.

However, if α is increased beyond a certain threshold value the iterates start to visit two points in turn. The period is said to have doubled and the system has undergone a bifurcation. As the parameter α is increased further still the number of points doubles again (period double by a pitchfork bifurcation). This continues until the number of points visited is infinite and the system behaves in a random fashion.

As α is increased further, however, a surprising thing happens. The system does *not* continue to become more chaotic. Instead the iterates start to behave in a quasi-periodic fashion with period

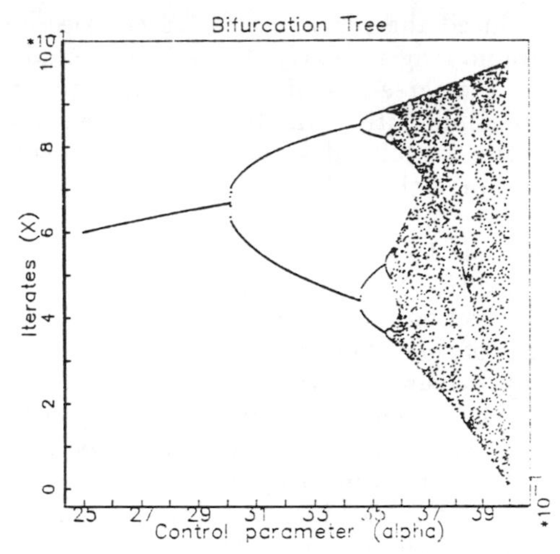

FIG.1 The evolution of the iterates as a function of α for the seed value $x = 0.65$

three (intermittency through tangent bifurcations). Figure 1 is an example graphical output from program 2 and shows a plot of the bifurcation tree. The pitchfork bifurcations can be clearly seen, and the band at $\alpha \approx 3.845$ shows the intermittancy, with the system visiting three points in turn.

Simple and spherical pendula

Programs 3 and 4 in the software package[1] are computer simulations of a driven, damped simple pendulum, and a driven damped spherical pendulum respectively. These can be used to study how chaotic motion can occur in the 'real world'. The students can in each case investigate the systems without damping, with damping, and then with both forcing and damping.

At the most simple level students can observe how the simple harmonic approximation for the simple pendulum fails when the initial angular displacement is large. The students are then introduced to the concept of phase space and how it relates to the behavior of the system, and to the existence of nodes or critical points that either attract or repel the phase trajectories. When damping is introduced the students can observe the existence of limit cycles, both stable and unstable, within phase space. When forcing is introduced the phase space trajectories cross over one another and are no longer clearly defined. The students can then obtain a plot of the three-dimensional phase space trajectories to separate out the ambiguities. In the case of the simple pendulum whose motion can be fully described by three parametric equations, each point on a three-dimensional phase space trajectory is unique.

It is also possible to study the Poincaré plot, which is analogous to the bifurcation tree. This shows, in the case of the simple pendulum, the position and angular velocity of the pendulum at periodic instants of time. When the pendulum is just set swinging with no forcing or damping, the equivalent Poincaré plot would be one point. Thus when the period of the system doubles the Poincaré plot displays twice as many points. See Fig. 2 for an example graphical output from program 3 which shows a plot of the strange attractor observed when the simple pendulum, subjected to certain values of damping and forcing, is behaving chaotically.

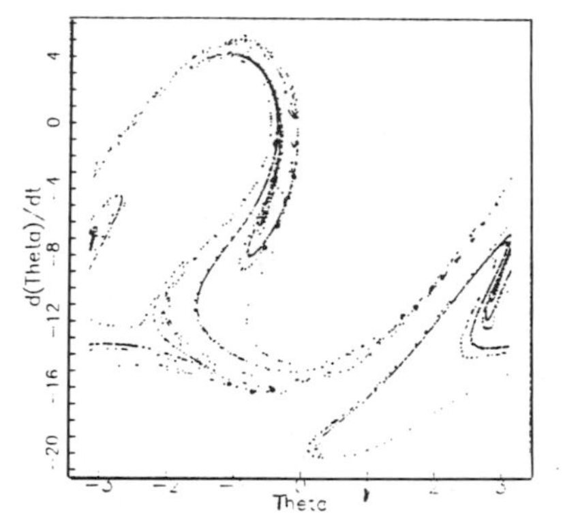

FIG.2 Simple attractor for the driven, damped simple pendulum

References

1 *Nonlinear Systems and Deterministic Chaos*, a software package and documentation available on request from The Software Development Unit (CTI), Department of Pure and Applied Physics, University of Salford, Salford, M5 4WT, UK

2 Predag Cvitanovic (ed.), *University in Chaos*, Adam Hilger, 1984

Multichannel analyser for student nuclear physics

R. BARLOW
Department of Physics, Manchester University, Manchester, UK

Abstract: The construction of a multichannel analyser at reasonably low cost for use in an undergraduate teaching laboratory is described, together with some applications and the reasoning behind our decision to build our own product.

The Nuclear Teaching Laboratory in Manchester provides experiments for second and third year students. Although third year students take ten laboratory days over an experiment and can thus get quite involved, second year students spend only two to three days, so that we have quite a short space of time to teach them a lot of physics that is unfamiliar to them – they have mostly done no nuclear and little atomic physics, particularly at the start of the year. This means we have to avoid getting them tied up in the technical details of the apparatus, yet at the same time give them plenty to look at, measure, and do. The multichannel analyser (MCA) is a highly effective tool for these purposes. It can capture and display a useful spectrum in a few minutes, so students get a quick and complete picture of their data, and can be used to take many measurements in a reasonable time. So when we wanted to expand and improve the nuclear experiments in Manchester, more MCA units were a high priority.

Unfortunately the cost of commercial units ruled out their purchase in large numbers. We were bitterly aware of the pitfalls of attempting to save money by manufacturing copies of commercial products in-house, knowing that such projects usually end in small savings and long delays, but satisfied ourselves that this case was different. Most of the above cost is due to the data storage and display hardware, which can now be handled by a cheap microcomputer. By building the ADC unit ourselves, and using a BBC microcomputer for the memory and display, we decided we could build a unit at a fraction of the commercial price.

Chips are available to turn a voltage into a binary number, and we set out to design a unit round one of these. A prototype was built that worked well on a pulse generator, but then failed to work with real pulses. The reason turned out to be that real pulses arrive at irregular intervals. When a pulse has been processed, the ADC input is then re-opened to receive another one; if this happens in the middle of a pulse on the input then misleading readings are obtained. This does not matter with regular pulses from a generator, or for some applications (like typical α experiments) where the rates are low and the probability for such an event small, but to cope with β and γ sources at reasonable rates, we discovered the hard way that some very involved input circuitry was necessary. Judging that to design this ourselves would not be a sensible use of resources, we adopted a design based on the Daresbury Laboratory's EC643 unit.

This is a high speed spectroscopy ADC, with a $3\,\mu$s conversion time, and a 'sliding scale' to eliminate non-linearity (a variable voltage, corresponding to between 0 and 63 channels, is added to the pulse at the analogue stage, and then subtracted at the digital stage, to randomise systematic effects). Although designed for 12 bit (4k) readout, we only used 10 bits, which saves money on the ADC chip (an AD579), and gives ample precision for our sodium iodide counters.

The ADC interfaces to an ordinary BBC computer through the 1MHz bus; the computer stores and displays the information, and from the keyboard the student can control the display, read the contents of selected channels, and so on. The teaching laboratories in Manchester contain a large number of BBC computers so that students are familiar with them. These are connected via Econet, so the program can be loaded quickly and conveniently (though an emergency version of the program is available on tape so that the lab is not at the complete mercy of the occasional network hangup). The computers can be used for other purposes, such as fitting straight lines to energy calibrations, or Fermi–Kurie plots of beta decay spectra, when not being used to control an MCA.

The program consists of two parts. A small assembler routine handles communication over the 1 MHz bus, gets a data value from it if there is one, updates the storage, and displays the histogram as the data is acquired. Detailed timing tests have not been done, but the fastest data taking rate appears to be about 900 events per second. The limiting factor here is the time taken by the software in plotting the histogram, which, because it has to apply the horizontal and vertical scale factors and uses system subroutine calls to do the actual plotting, takes much longer than the data acquisition and storage. However this data rate is quite satisfactory for our purposes, and the sight of a historgram actually building up as the student watches is very important in showing them what is going on, so there are no proposals to change this.

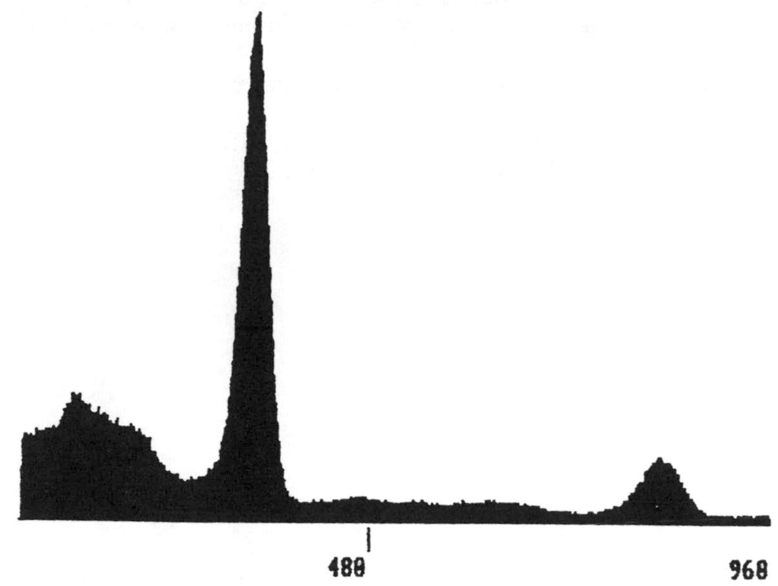

Fig.1 Typical spectrum of Na^{22} gamma-rays, obtained with the MCA. The peaks occur at 511 and 1275 keV

When a key is pressed this generates an interrupt and control passes to a BASIC program that communicates with the student. He or she can change vertical and horizontal scales, clear the data, restart data taking, ask for details of the contents or integrated contents of a range of channels, move a cursor around, save data to file and retrieve them, and print the display on a dot-matrix printer. It was found straightforward to provide all the facilities found on a conventional MCA, except that the BBC has no knowledge of the deadtime of the MCA, so a reading of the 'true' time (i.e., elapsed time less deadtime) is not possible. Saving data on files for later analysis is crucial in β-ray experiments, where the spectrum is processed to obtain a Fermi-Kurie plot and this facility is also used sometimes by students doing other experiments.

Seven units were built, at a cost of about $1 000 each (not including the computer, or labour). They are currently used for undergraduate experiments, including

- studying the Compton effect by examining $Cs^{137}\gamma-$rays scattered at different angles from a target
- measuring the energy loss and straggling of α-particles in argon at low pressures, and verifying the Bethe-Bloch formula
- studying the γ-ray spectra of simple sources, and the properties of γ-rays, their absorption and detection
- measuring the energy spectra of β decays and constructing Fermi–Kurie plots.

The units are easy to use and popular with the students, and have performed reliably over many months.

Inexpensive 2000–channel analyser with applications in nuclear science

A.J. Woods and M.J.W. Elliott
Electronics, Physics and Electrical Engineering Unit, School of Engineering, Oxford Polytechnic, Oxford, UK

Abstract: A 2000 channel multichannel analyser (MCA) has been designed and constructed for use in an undergraduate nuclear science laboratory. The MCA is controlled by a BBC microcomputer. Software has been written to facilitate most of the usual functions of expensive commercial instruments. An experiment on beta decay illustrates one application.

Multichannel analysers (MCAs) using pulse height analysis (PHA) or multichannel scaling (MCS) are essential in all modern undergraduate nuclear science courses. Although no longer totally exotic pieces of equipment (as pointed out by Chomet et al. [1]), the average cost of even a small portable commercial MCA is several thousand pounds. In the present financial climate this can prove prohibitive. The availability of even a basic microcomputer (e.g., BBC) enables a 2 000-channel MCE to be built for only a few hundred pounds. This can make a considerable difference when wishing to run experiments on alpha, beta, gamma, X-ray, and Mössbauer spectroscopy, and so on, concurrently.

Figure 1 shows a block diagram of the system developed and used in the nuclear science laboratory of Oxford Polytechnic. Signals from a detector/amplifier combination are processed by the MCA and then passed to a host microcomputer for storage and display. When PHA is used, the MCE processes a pulse in 30 μs and then signals the microcomputer that data are ready for transfer. Parallel-processing techniques are used enabling the MCA to process a second pulse while the microcomputer stores the first. This gives a theoretical maximum count rate of about 30 000 cps with a suitable microcomputer.

Figure 2 shows a block diagram of the MCA. Because of the random nature of radioactivity, pulses entering the MCA need to be detected and trapped using an ultra-high speed peak detector.

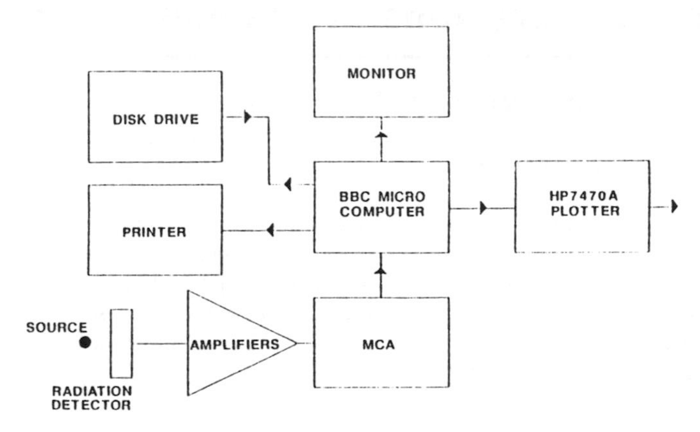

Fig.1 Typical system configuration

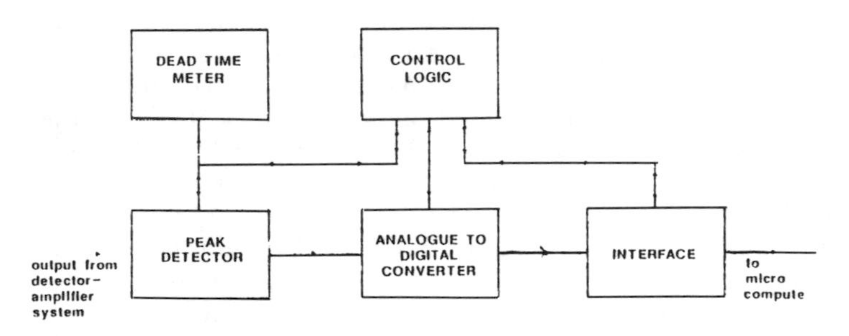

Fig.2 MCA block diagram

The output from this is stored using a sample and hold amplifier. The peak height is then converted into an 11 bit number using the ubiquitous AD574 analogue to digital converter. This number is stored in a single count memory to allow parallel processing to take place. Control and interfacing logic completes the MCA hardware.

The system utilises a BBC microcomputer with 20K memory extension, disk-drive, monitor, printer and plotter. Spectra can be stored and retrieved from disk; hardcopy is in the form of channel listings or graphs. Versatile software making extensive use of procedures has been developed for the MCA using BBC-BASIC. 80% of the monitor screen is reserved for displaying data. A text window appears at the top of the screen and is used to display menus of available options and control communication with the operator.

The BBC's function keys are used heavily in conjunction with the menus to control the MCA. The data display rescales if the total count in any one channel exceeds 90% of the scale maximum. Data are displayed in two formats: a pulse height spectrum of all 2 000 channels or a smaller spectrum showing a 'page' of 400 channels. The operator is able to switch between display modes, increment page number (i.e., move the start of the display up by 100 channels), decrement page number, select new page, change the scale on the display, and manipulate the cursor. The function keys and menus are also used to start and stop data acquisition, print, plot, store, and retrieve data and also to execute the subtraction and Kurie plot routines.

An experiment in beta spectroscopy will serve to illustrate an application of the MCA. An article on beta spectroscopy by Chomet et al.[1] in an earlier volume of this series described an interesting method for 'creating' a beta spectrum within the phosphor of a scintillation counter. The phosphor is irradiated with neutrons that convert iodine-127 into beta radioactive iodine-128. The counter output pulse height spectrum is subsequently converted into a Kurie plot for end-point energy determination. In the present experiment conventional sources are used with a semiconductor detector. A suitable detector is a Princeton Gamma-Tech silicon surface barrier detector (Model PD-50-24-1000M) which has a depletion depth of 1 000 μm and operates with a bias voltage of +350. Figure 3 shows the system. In order to provide an energy calibration a bismuth-207 electron conversion source is used. 'Monoenergetic' electrons at 482, 554, 972, 1044, and other energies

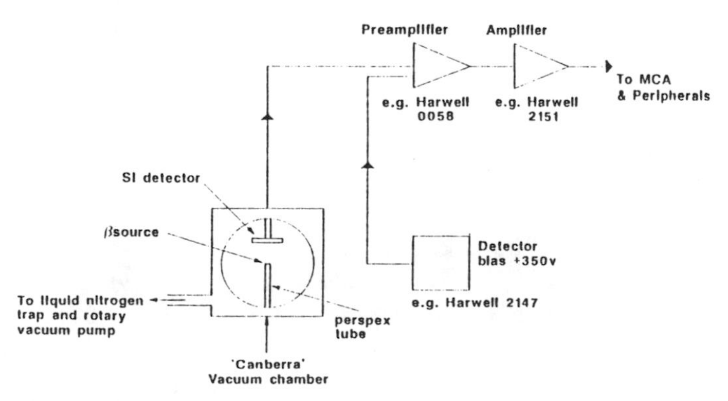

Fig.3 Beta spectrometer

up to 2400 keV are emitted. Range-energy graphs for electrons in silicon[2] show that 800-keV electrons have a range of $rm1\,000\mu m$. However, owing to the zig-zag nature of the path of beta particles in the depletion layer, peaks for the four lower energies are easily obtained.

A plot from the MCA plotter with the corresponding energy-channel calibration is shown in Fig. 4. Thallium-204 is a useful source for study of a normal continuous beta spectrum. With a

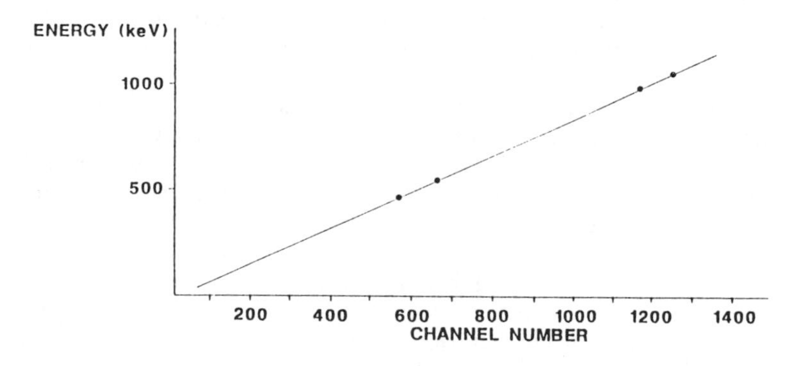

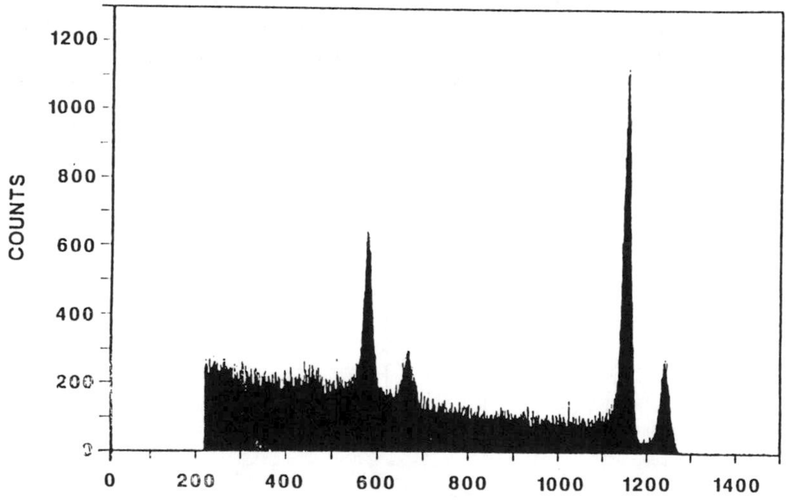

FIG.4 HP7470A plotter output for energy calibration with bismuth 207 internal conversion electrons

maximum energy of about 764 keV the pulse height spectrum falls within the energy calibration range. The data are converted into a Kurie plot by the software. The operator enters energy-channel data from the bismuth-207 plot and the MCA calibrates itself using a least squares fit on this data.

The software then produces a Kurie plot of the thallium data. The beta spectrometer in use here measures the beta particle energy T directly whereas in a magnetic deflection spectrometer the beta particle momentum distribution is measured. In terms of energy the Kurie plot for allowed transitions can take the form suggested by Fabry et al.[3], that is plot

$$\frac{N(T)}{F(1+\epsilon)(\epsilon^2+2\epsilon)^{1/2}}$$

against channel number, where $N(T)$ is the number of counts in a given channel, $\epsilon = T/m_0c^2$, $m_0c^2 = 511\,\mathrm{keV}$ is the electron rest energy, and F is the Fermi function.

Calculation of the Fermi function or interpolation from tabulated data can be very tedious. Recently, Venkataramaiah et al.[4] have shown that F is easily estimated from the empirical expression $F = (A + B/\epsilon)^{1/2}$ where A and B are constants for a given beta emitter. Values of A and B are listed for many common beta emitters in Ref. 4. For thallium-204, $A = 161.94$ and $B = 403.58$. Strictly, the beta decay of thallium-204 is not an allowed transition and a shape factor should be introduced into the Kurie plot. However this makes very little difference to the high energy end of the

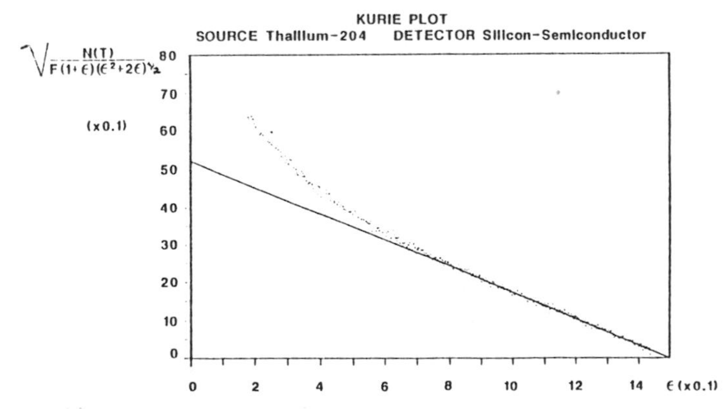

FIG.5 HP7470A plotter output of Kurie plot

spectrum and is not worth introducing when the main concern is to deduce the end-point energy. The software does this by a least squares fit on the last 40% of the thallium Kurie plot and uses it to calculate the intercept on the T axis. The value obtained for the end-point energy for this is within 0.25% of the literature value. Further details can be obtained from the authors.

References

1 S. Chomet, M.A. Hill, and D.P. Lidiard, *Physics Experiments and Projects for Students*, vol.1, 173 (1985)

2 S. Deme, *Semiconductor Detectors for Nuclear Radiation*, chapter 1, Adam Hilger, (1971)

3 M. Fabry, J.R. Cussenot, and A. Picot, *European J. Phys.* **2**, 129 (1981)

4 P. Venkataramaiah et al., *J. Phys. G. Nuclear Phys.* **11**, 359 (1985)

Microcomputer-based alpha-particle spectrometer

R.J. BISHOP
School of Applied Physics, Kingston Polytechnic, Kingston upon Thames, Surrey. KT1 2EE, UK

Abstract: A simple, inexpensive alpha-particle energy spectrometer based on the BBC microcomputer is described. The system is relatively easy to construct and operate and should prove very useful in the teaching of nuclear radiation science at sixth form and undergraduate level.

System hardware

The basic block diagram of the hardware used is shown in Fig. 1. The signals from the detector are first amplified and then fed, via a suitable interface, to a standard BBC microcomputer configured to act as a 256-channel multi-channel analyser.

The charge sensitive amplifier produces output pulses of amplitude about 100 mV when used with natural alpha-particle emitting radiation sources. The voltage amplifier provides additional gain of a factor of 20 and brings the pulse amplitudes up to the operating range of the following analogue-to-digital converter (0 – 2.5 V). The comparator is triggered by input signals exceeding its threshold and in turn triggers a monostable.

The monostable positive output pulse is used as the logic input to the sample and hold circuit. This latter circuit has as its analogue input the output from the voltage amplifier. Whilst the logic input is high the output of the sample and hold circuit follows the analogue input signal. When the logic input is taken low the output of the sample and hold circuit is maintained at that value of the analogue input existing at the instant the logic input switches.

This 'held' voltage, which is proportional to the radiation energy deposited in the detector, is applied to the analogue-to-digital converter (ADC). The negative output pulse from the monstable acts as the convert command signal to the ADC. The converter is cleared by a low-going convert command signal and begins its conversion when the signal goes high again. The ADC then rapidly (9 μs) converts the analogue input signal into a digital output that is subsequently fed into the microcomputer.

The monostable negative output pulse is also used as an interrupt signal to the microcomputer. This signal causes the computer to temporarily stop doing its present task and to begin a routine that reads into the computer the digital output from the ADC. The ADC output is enabled upon receipt of a signal from the computer indicating it is ready to receive data.

Figure 2 shows the circuit diagram of the complete electronic system except for offset null adjustment circuitry which has been omitted from the two operational amplifiers for reasons of clarity. All the components are readily available commercially.

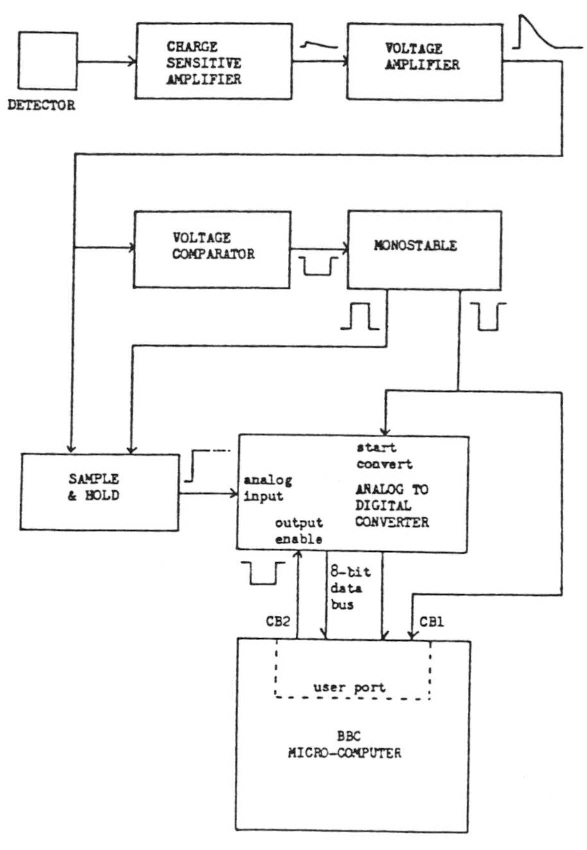

Fig.1 Block diagram of the electronic system

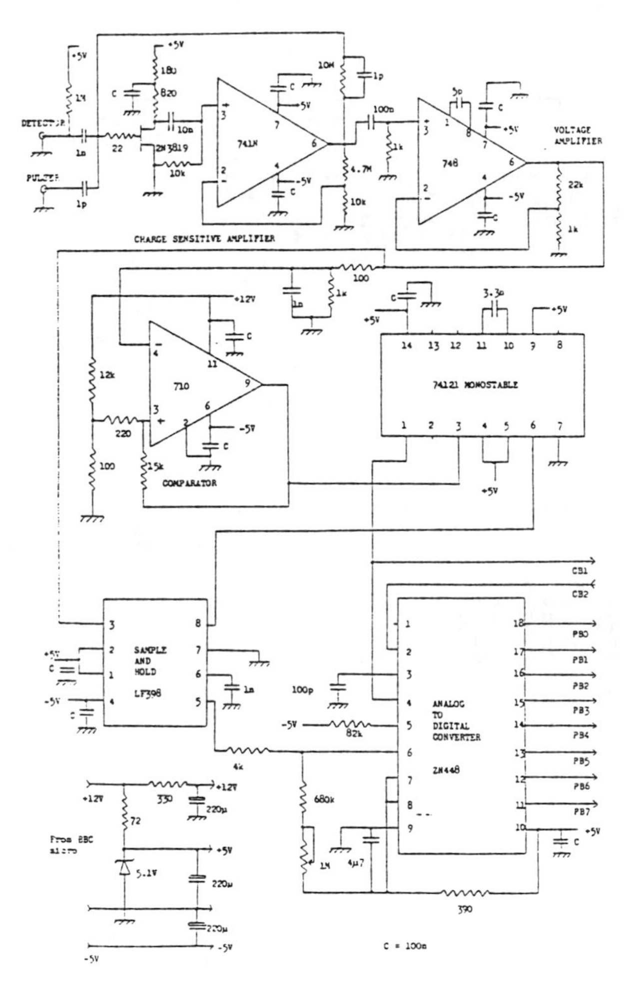

FIG.2 Circuit diagram of the complete electronic system

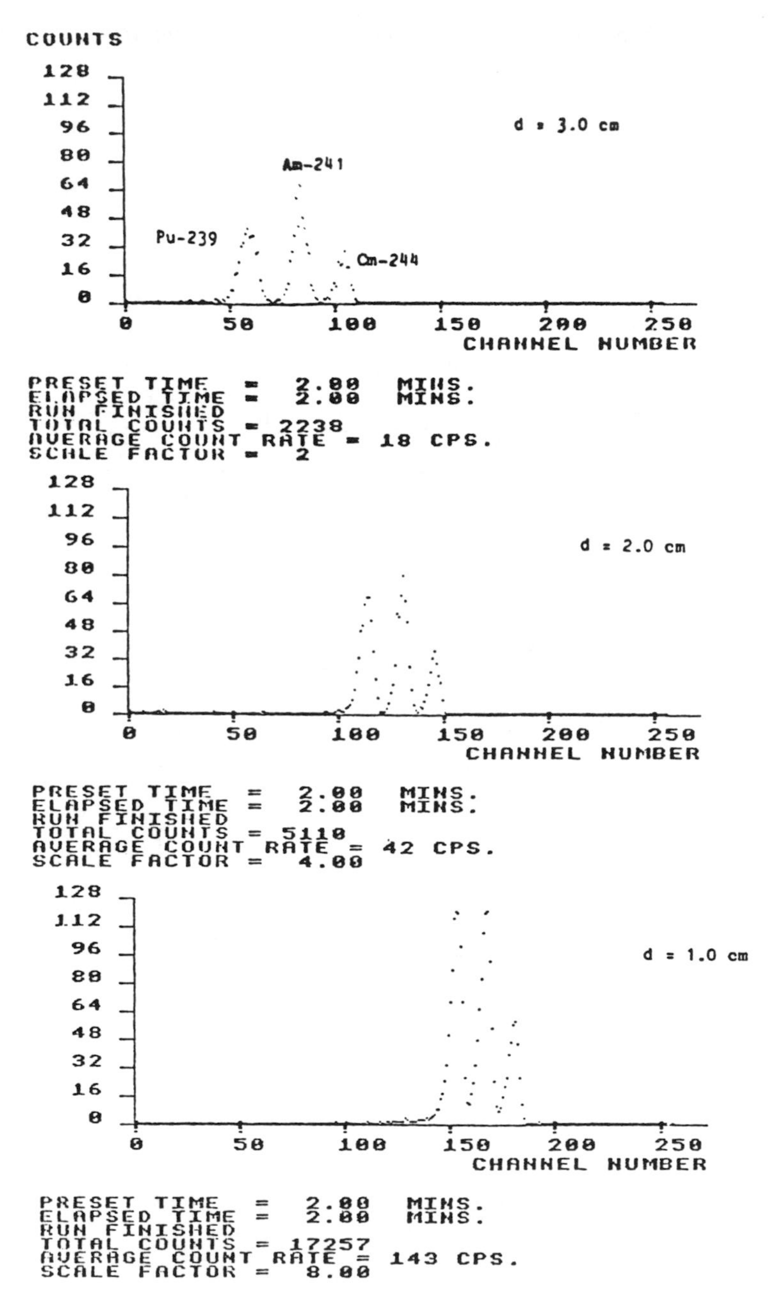

FIG.3 Energy spectra from a mixed alpha-particle source as a function of source-detector separation (d) in air

System software

The software used must, as a minimum, be responsible for the setting up of the user port on the microcomputer, the input of data from the ADC via the port and the storage of that data in appropriate memory locations, and for the actual display of the stored data in a user-friendly format on the VDU screen.

Additional software features included by the author are automatic and manual scaling of the stored data prior to display, ability to preset the count time, peak integration, energy calibration, printout of data, screen dump, and storage and retrieval of data on floppy discs. Direct screen addressing is used to give a rapid display update in real time.

Copies of the software are available from the author.

System performance

As an example of measurements made with this system, Fig3 shows energy spectra from a mixed alpha-particle emitting source as a function of source-detector separation in air. By substitution of other suitable nuclear radiation detectors (plus their associated preamplifiers and power supply units), such as proportional and scintillation counters and Si(Li) or Ge semiconductor detectors in place of the alpha detector and its preamplifier, the system can be used for energy measurements on other types of nuclear radiations.

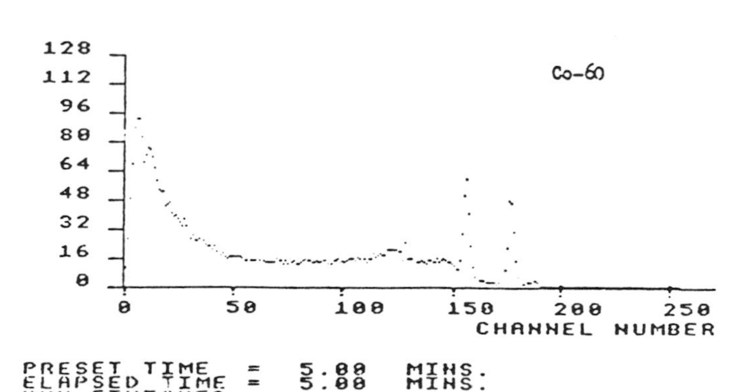

FIG.4 Gamma-ray spectrum from a Co-60 source

Figure 4 shows the gamma-ray spectrum of Co-60 measured with a (somewhat ancient) Ge (Li) semiconductor detector.

The maximum counting rate of the system before onset of degradation of energy resolution is about 2000 c.p.s. and is limited by the present software. Each of the 256 channels has a capacity of 64k counts but this can easily be increased to 16M counts if needed. Linearity was measured to be about 1%. The total cost of the electronic hardware of Fig. 2 is less than $45.

Clearly, the performance of the system is somewhat limited compared with (very expensive) commercial systems, but it is adequate for its intended use.

Muon lifetime measurement

M.G. GREEN
Royal Holloway and Bedford New College, University of London, Egham, Surrey, UK

Abstract: The muon lifetime is one of the few properties of elementary particles that can be measured in the undergraduate laboratory. Cosmic rays provide the source of the muons since some will stop in a suitable detector, for example a large block of scintillator. These decay a few microseconds later thus producing two pulses from a photomultiplier attached to the scintillator. The time delay between these pulses is measured and passed on to a microcomputer which produces a histogram from many such decays. A least squares fit in the computer produces a value for the mean lifetime of the muon.

The muon (μ) is an elementary particle very similar to the electron, but about 200 times heavier. It was discovered in the 1930s in sea level cosmic-rays. Muons are produced when primary cosmic-rays, principally protons, hit the upper levels of the earth's atmosphere and produce pions. The pions quickly decay to muons which make their way down to sea-level decaying as they do so to electrons and neutrinos (ν). Thus at sea-level, cosmic rays are a mixture of undecayed muons, electrons and neutrinos, the latter being almost undetectable since they can pass straight through the whole earth with negligible probability of interacting.

$$p + \text{nucleus} \rightarrow \pi + \ldots$$

$$\pi \rightarrow \mu + \nu$$

$$\mu \rightarrow e + \nu + \bar{\nu}$$

The process that causes the muon to decay is the same one that is involved in the beta decay of radioactive elements; indeed the muon is the lightest particle that undergoes beta decay.

The measurement of the muon lifetime is one of the few experiments in elementary particle that can be carried out in the teaching laboratory and refinements to it have been described in several papers[1–4] since it was first suggested by Melissinos[5] in 1966. In the earliest versions just a few decays were observed, but advances in electronics have allowed significant improvements. The ubiquitous microcomputer also has an obvious application in such an experiment and hence in recent years we have been able to collect tens of thousands of examples of decays and analyse them. As a result systematic errors, in particular the effect of muon capture by nuclei, now limit the accuracy of the lifetime measurement.

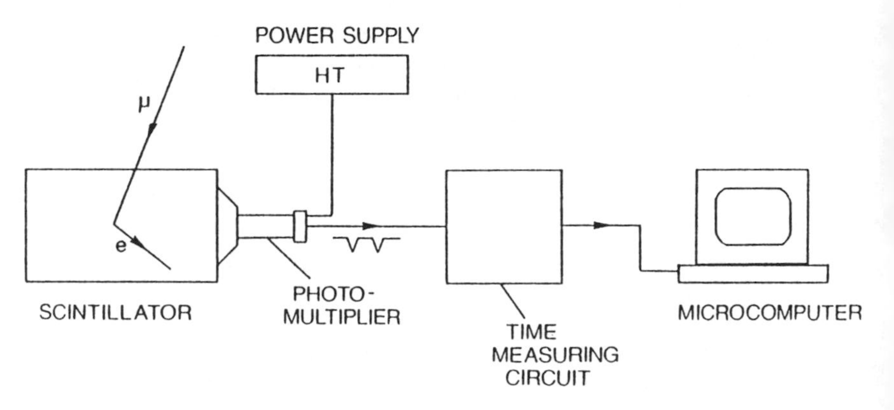

Fig.1 Block diagram of the apparatus

The version of the experiment described here was developed and built as a project by two third-year students. A block diagram of the apparatus is shown in Fig. 1. The detector is a large piece of scintillator, $0.48\,\mathrm{m} \times 0.18\,\mathrm{m} \times 0.12\,\mathrm{m}$, contained in a light-tight box. When muons enter the scintillator, light is emitted and is detected by the photomultiplier that gives out an electronic pulse. A few of the muons stop in the scintillator and decay a short time later emitting an electron that gives a second flash of light. Since the light flashes last just a few nanoseconds (ns), while the mean lifetime of each muon is about 2.2 microseconds (μs), the two pulses are easily resolved. They are fed to a simple electronic circuit that measures the time delay between the two to an accuracy of $0.1\,\mu$s and sends the value of this delay on to the computer (Fig. 2).

The electronic circuit works as follows. A photomultiplier pulse above the discriminator threshold sets the counter enable to count. Clock pulses from the 10 MHz clock are then counted until a second input pulse arrives or the counter reaches terminal count. A second pulse stops the counter with the time interval between the

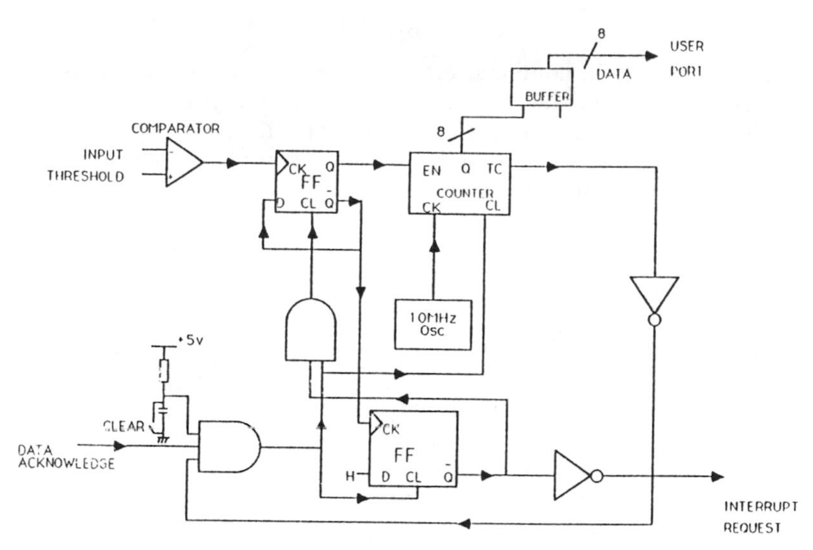

FIG.2 Block diagram of the circuit, including connections to the microcomputer

pulses buffered on to the data output lines and sets the interrupt request line to the computer. A data acknowledge response from the computer clears and resets the system. A terminal count stops, clears, and resets the system. The switch can be used to disable the system.

A small routine in the computer written in assembler handles the interrupts and passes the time measured in each decay to a program written in BASIC which collects it as a histogram. The histogram can be displayed, analysed, or saved to floppy disk. Figure 3 shows the results from a 72 hour run containing about 25,000 decays. Since the decay process is random the expected exponential distribution is clearly seen (as in radioactive decay, we do not need to know when the muon was born but can start timing at any point during its life). In addition there is a small, constant background that can be clearly seen on the log plot.

The presence of the background under the exponential complicates the analysis of the data. The simplest way of dealing with it is to measure it in the high t region and then remove it by subtraction from under the exponential. However, we have also developed a second approach, namely to fit the whole distribution to the function

$$N(t) = n_0 e^{-t/r} + c$$

using an iterative, nonlinear least squares fit.[6] Although somewhat more complicated this method has the advantages of giving reliable errors on n_0, λ, and c and a χ^2 for the fit. It also has the educational advantage of being a good example of the nonlinear least squares technique.

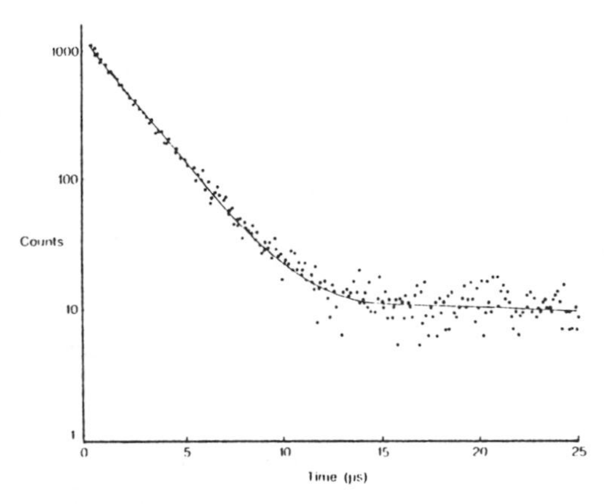

FIG.3 Log plot of data from a 72 hour run together with a fit to the sum of exponential plus a constant background

The fit to the data of Fig. 3 gives a mean muon lifetime τ_μ of $2.121 \pm 0.017\,\mu s$ and has a χ^2 of 234 for 244 degrees of freedom. This lifetime value is 4.5 standard deviations low compared with the known value of $2.198\,\mu$s as a result of nuclear capture of some of the μ^- in the carbon nuclei of the scintillator. In this process a muon is trapped in a high atomic orbit and rapidly cascades down to the K-shell displacing electrons. Since the muon is some 200 times heavier than the electron, its Bohr radius, which is proportional to $1/Zm$, is small enough for it to move very close to the nucleus, particularly for high Z nuclei. As a result the rate for the capture process

$$\mu^- + X \rightarrow \nu_\mu + X'$$

is significant compared to decay for the carbon nuclei in plastic scintillator.

In a cost-saving variation of the experiment we used a 25 ℓ drum of liquid scintillator as the detector. The scintillator was produced by dissolving 100 gm of PPO and 1.25 gm of POPOP in toluene. This version had essentially the same counting rate as the plastic scintillator and produced comparable results for the muon lifetime but it brings with it the safety hazard of the use of toluene and does not reduce the muon capture problem.

Acknowledgments

I acknowledge the help and advice of my colleague J.A. Strong over the years of development of this experiment, particularly with the electronics. The two third-year physics students who built the version described here were Christina Strunks and Julian Potter.

References

1 A.E. Hall, D.A. Lind and R.A. Ristinen, *Am. J. Phys.*, **38**, 1196 (1970)
2 A. Owens and A.E. Macgregor, *Amer. J. Phys.* , **46**, 859 (1978)
3 R.J. Lewis, *Am. J .Phys.* , **50**, 894 (1982)
4 T.Ward et al., *Am. J .Phys.* , **53**, 542 (1985)
5 A.C. Melissinos, *Experiments in Modern Physics*, Academic Press, p.411
6 M.G. Green, J.M. Potter, and J.A. Strong, *submitted to Amer. J. Phys*

Appendix: Chaos in the laboratory

```
REM Chaos in simulated varactor diode/inductor circuit
REM M.R.Halse & S.J.Rogers, University of Kent, Jan88
REM ATARI ST FAST BASIC
    GRAFRECT 0,0,640,400:CLG 0:MARKCOL 1
REM
REM Vd,Id=diode voltage,current, Vo=diode threshold voltage
REM DEFINE UNITS
                mH=1E-3:pF=1E-12:volts=1:nanosecs=1E-9:ohms=1
REM set component values and drive voltage
          R=60*ohms:L=2.5*mH:C0=33*pF:Vo=0.6*volts:Vmax=1*volts
REM set frequency
                w=2*PI*500000:period=1/500000:dT=period/250
REM set iteration parameters
goflag=0:I1=0:I2=0:T=0:E=dT*dT/L:F=0.5*R*dT/L:G=1E-5
REM main program starts here
    HIDEMOUSE:PRINT "Vmax   ";Vmax:PRINT "w      ";w
    REPEAT
        PROCiterate
        PLOT 320-50*V,190+150000*I2
        PROCkey
    UNTIL goflag=1
REM now plot chaos bifurcation diagram
    LINE 20,0 TO 20,400:LINE 0,30 TO 640,30
    settletime=20*period:plottime=50*period
    FOR Vmax=4E-3 TO 3 STEP 4E-3
      CLS:T=0:PRINT "Vmax   ";Vmax:PRINT "w      ";w
      REPEAT
          PROCiterate
50        IF (ABS(cos)<0.013 AND T>settletime AND sin>0) THEN
                 PLOT 20+Vmax*125,30+300000*I2
          ENDIF
      UNTIL T>(settletime+plottime)
    NEXT Vmax
PRINT "finish":X=GET:END
-------------------------------------------------------------
DEF PROCiterate
           wT=w*T:sin=SIN(wT):cos=COS(wT):V=Vmax*cos
           Vd=V-L*(I2-I1)/dT-R*I2
           IF Vd>0  THEN
                         C=C0*(1+100*Vd*Vd)
                   ELSE C=C0*0.8944/SQR(0.8-Vd)
           ENDIF
           IF Vd>Vo THEN
                         Id=Vd*G+1.5*(Vd-Vo)*(Vd-Vo)*(Vd-Vo)
                   ELSE Id=Vd*G
           ENDIF
100        I3=(2*I2-I1-E*(w*Vmax*sin+(I2-Id)/C)+F*I1)/(1+F)
           I1=I2:I2=I3:T=T+dT
ENDPROC
-------------------------------------------------------------
DEF PROCkey
   key%=INKEY
   IF key%=-1         THEN ENDPROC
   IF key%=ASC(" ")   THEN CLG 0
   IF key%=ASC("V")   THEN Vmax=Vmax+0.1
   IF key%=ASC("v")   THEN Vmax=Vmax-0.1
   IF key%=ASC("W")   THEN w=w+100000:dT=2*PI/w/250:E=dT*dT/L:F=0.5*R*dT/L
   IF key%=ASC("w")   THEN w=w-100000:dT=2*PI/w/250:E=dT*dT/L:F=0.5*R*dT/L
   IF key%=ASC("g")   THEN goflag=1
   CLS:CLG 0:PRINT "Vmax   ";Vmax:PRINT "w      ";w
ENDPROC
-------------------------------------------------------------
```